# ELECTROOPTICS
# and
# CONDUCTOMETRY
# of
# POLYDISPERSE
# SYSTEMS

# ELECTROOPTICS and CONDUCTOMETRY of POLYDISPERSE SYSTEMS

Anatoli A. Trusov
Vladislav V. Vojtylov

CRC Press
Boca Raton Ann Arbor London Tokyo

**Library of Congress Cataloging-in-Publication Data**

Voĭtylov, V. V. (Vladislav Viktorovich)
[Ėlektrooptika i konduktometriia polidispersnykh sistem. English]
Electrooptics and conductometry of polydisperse systems / Anatoliĭ A. Trusov and Vladislav V. Voĭtylov.
p. cm.
Authors' names in reverse order in the original Russian ed.
Includes bibliographical references and index.
ISBN 0-8493-7730-7
1. Electrooptics. 2. Colloids—Electrical properties. 3. Electrochemistry. I. Trusov, A.A. (Anatoliĭ Anatol'evich) II. Title.
QC673.V65 1993
541.3′45—dc20 93-22827
CIP

SMART Publishing, Ltd., Ioffe Physico-Technical Institute, St. Petersburg, Russian Federation. Editing, formatting with TEX and a camera-ready copy were provided by the Editorial Office at the Ioffe Institute, 26 Polytechnicheskaya st., 194021 St. Petersburg, Russian Federation.



Direct all inquiries to CRC Press, Inc., 2000 Corporate Blvd., N.W., Boca Raton, Florida 33431.



International Standard Book Number 0-8493-7730-7

Library of Congress Card Number 93-22827
Printed in the United States of America 1 2 3 4 5 6 7 8 9 0
Printed on acid-free paper

# Contents

# INTRODUCTION

The use of electric fields in the study of colloidal dispersions has long attracted the researchers' attention. In 1809 F. F. Reuss discovered electrophoresis, and in 1875 D. Kerr was the first to observe an electrooptic effect. D. K. Maxwell pioneered conductometric studies of dispersions. The further development of the science of colloidal systems showed that these phenomena were interrelated and their manifestation was determined by the properties of an electrical double layer at the separation boundary between the solid and the fluid. Although the evolution of ideas concerning electrokinetic, electrooptic and conductometric phenomena in colloidal dispersions has been fairly irregular, at present the experimental techniques based on their application play a leading role in colloidal research.

Electrokinetic experiments were initiated at the beginning of the 20th century and have provided important qualitative and semi-quantitative data on the electric charge and structure of a particle double layer. They still remain a major source of information about electrical properties of dispersed material.

Systematic electrooptic studies of colloidal solutions were initiated by N. A. Tolstoi. These studies were summarized in the work by N. A. Tolstoi and P. P. Feofilov (1949). It was shown that the electrooptic effect which was observed in a Kerr cell filled with a colloidal solution was associated with dichroism rather than birefringence, the latter being only a weak side effect accompanying dichroism. This conclusion became an underlying concept in the development of electrooptic methods in this country. In his studies of dichroism in electric fields of rectangular pulses and rotating fields, N. A. Tolstoi found that colloidal particles had a permanent dipole moment. Soon this observation was confirmed by C. T. O'Konski, who studied electric birefringence in colloidal suspensions. Further research into the nature of the permanent dipole moment showed that it was determined by groups of polar molecules adsorbed on the particle's surface.

During the 1960s the development of dispersion electrooptics went on very intensively. The Bulgarian scientists A. Sheludko and S. P. Stoylov and the British researchers B. R. Jennings and H. J. Gerard started systematic investigations on the light scattering by colloidal solutions in an external electric field. Their efforts were focused on the understanding of the field-induced moment of particles and anisotropy of their polarizability in biological dispersions. Steady-state electrooptic phenomena and transient processes, that arise when an orienting electric field is turned on or turned off, were studied in uniform sine-shaped fields. However, the values of polarizability anisotropy obtained in different ways differed by a factor of several dozens. Later it became clear that this difference was due to the neglect of particle size variation in real polydisperse systems. The development and testing of polydispersity theory provided eventually a good agreement between the results of experimental data processing.

Of great importance in electrooptic studies is an understanding of the mechanism of particle orientation induced by the applied electric field. This mechanism became clear when the thermodynamic theory of non-equilibrium processes was applied to the description of the moments of force exerted on a particle in an electric field. It follows from this theory that the particle polarization and the moments of force are greatly affected by the surface conductivity of a particle associated with its electrical double layer. The combination of electrooptic theory for polydisperse systems and thermodynamic description of polarizability anisotropy for non-spherical particles have made it possible to study adequately the double layer characteristics using electrooptic methods.

The development of a theory for description of an electrically induced dipole moment of a colloidal particle raised the significance of conductometric data in the study of polarization mechanisms of the double layer. The electrical conductivity of dilute and concentrated dispersions was analyzed theoretically and studied experimentally.

In recent years conductometric methods have been extended to cover such phenomena as conductivity anisotropy induced by an external electric field orienting non-spherical particles. These methods provide adequate information on the particle polarizability in polydisperse systems and on the dynamics of electrical interactions of particles.

Combined investigations on conductivity anisotropy and electrooptic phenomena have revealed a common mechanism of particle orientation, which can be profitably used for studying the properties of the double layer of particles in polydisperse systems.

In this book we do not try to review all electrooptic studies but focus on particle orientation by alternating uniform electric fields and its relaxation. These studies are concerned with the anisotropy of particle polarizability in polydisperse colloidal systems. The description of optic effects in dispersions and the theory of the double layer structure are presented only to the extent necessary for a discussion of electrooptic and orientational conductometric methods and their potentialities.

The book consists of six chapters. In Chapter 1 we discuss the basic optical properties of dispersions and approaches to the problems of light scattering by a particle. The formulation of a typical electrooptic problem is presented. Chapter 2 describes electrooptic effects in polydisperse systems and the ways of determining polydispersity. The theory of the electrical double layer and particle polarization necessary for the further presentation of electrooptic and conductometric data is given in Chapter 3. In Chapter 4 we analyze the monomorphic approximation and its application in electrooptic investigations. Some research data on the anisotropy of polarizability for several types of polydisperse systems are presented. Chapters 5 and 6 are concerned with electrical conductivity of polydisperse systems. Measurements of conductivity anisotropy in an electric field are used to study the double layer characteristics in several types of dispersion. We also describe a way of finding particle sizes from conductivity anisotropy measurements.

# Chapter 1

# OPTICAL AND ELECTROOPTIC PROPERTIES OF DISPERSIONS

## 1.1. OPTICAL PROPERTIES OF DISPERSIONS AND LIGHT SCATTERING THEORY

Optical methods have proved very effective in the study of dispersions. Sometimes, valuable qualitative and semi-quantitative information can be obtained with the naked eye, using the sun or a lamp as a light source. For instance, if a dilute dispersion has a bluish color, this implies that all or most of the particles have sizes comparable with the blue light wavelength and are close to 0.3–0.4 $\mu$m. By a dilute dispersion we mean a system with interparticle spacings at least one order of magnitude larger than the particle sizes.

Unlike dilute dispersions, a concentrated dispersion has the color of dispersed material. The same is true for systems containing particles, whose size is much larger than the incident wavelength. For daylight the particle sizes fall within a few micrometers. Evidently, size variation of light scatterers must change the color of a dispersion. This effect is observed when particles coalesce to form aggregates, because in that case the light is scattered by an aggregate rather than by a single particle.

Optical properties of dispersions are associated with light absorption and light scattering, as well as with the refractive index difference between the dispersion and the dispersive medium. In real dispersions, optical phenomena may occur concurrently. Light absorption by particles is always accompanied by its scattering. For example, even strongly light-absorbing graphite scatters about 30% of the incident light in aqueous media.

All observable optical phenomena are associated with polarization of the electron shells of the atoms and molecules comprising a particle and with the dipole moment induced by the incident wave electric field. The induced dipole moment oscillates at a frequency of about $10^{14}$ Hz and produces a new electromagnetic wave. The particle with an induced dipole moment scatters the light, partly absorbing it. It acts as a dissipative oscillating system with an oscillation phase different from that of the incident wave electric field.

Theory of optical phenomena in colloidal dispersions is based primarily on a spatial and temporal description of the light-induced dipole moment of a colloidal particle. Such a description is valid, however, only for particles with a size much smaller than the incident wavelength. But if the sizes fall within a range from several hundredths of a micrometer to a few micrometers, a rigorous mathematical description of a dipole moment represents an extremely complicated problem, which becomes virtually unsolvable for particles of irregular shape. The value of a dipole moment depends on the particle orientation relative to the electric field vector, so a dispersion with orientationally ordered particles is optically anisotropic. Generally, orientational order can be created by an external electric field with a frequency several orders of magnitude lower than the light field frequency. Then the induced optical anisotropy can be related to the appearance of birefringence or to the intensity variation of the light scattered or absorbed by particles. Since particle orientation is determined by the interaction between the dipole moments and the electric field inducing the orientation, observable optical anisotropy can provide information about the dipole moments and polarizability of particles. Finding a correlation between the characteristics of induced optical anisotropy and the electrical properties of particles is a very important aspect in the study of electrooptic behavior of dispersions. Another objective of electrooptic studies is determination of the sizes and shapes of particles. This information can be obtained from investigation of unsteady-state electrooptic effects.

Since most of the incident light is scattered, valuable data can be obtained from experiments on electrooptic phenomena associated with scattering anisotropy. On the other hand, some electrooptic effects that are unrelated to light scattering by particles may be screened from observation by certain features of the scattering. For example, dichroism associated with light scattering anisotropy considerably complicates observation of birefringence and intrinsic dichroism of colloidal particles.[1] Thus, a rigorous theoretical description of optical and electrical effects in dispersions directly involves the theory of light scattering by matter.

The theory of light scattering has a long history, but even today it is an actively developing area of optics. Various aspects of this theory have been discussed in a number of monographs.[2–5] A fairly complete review of available literature on light scattering by particles has been made by N. J. Khlebtsov.[6] Applied research data on light scattering are reported in periodicals.[4,7–9] Besides, there are some books, in which the available information has been systematized and presented quite comprehensively (see, for example, References 3–5,10).

The theory of light scattering should be regarded as an important area in the electrodynamics of inhomogeneous media, which considers scattering as radiation transfer. No doubt, this is true only if the intensity of an incident light beam is not too high and the electromagnetic fields affect but slightly the scattering material. The basic parameter in the theory of radiation transfer is energy, a magnitude proportional to the squared electric field strength of a light wave. This approach allows one to deal with photometric values that can be measured in experiment. Photometric values are commonly used in studies on light scattering by particles. The transfer theory does not consider collective scattering phenomena but may take into account multiple scattering.

Other approaches to the description of light scattering by particles consider single scattering events and are applicable only to dilute dispersions. In this case the problem is reduced to a theoretical interpretation of scattering by a single particle.

Much theoretical work has been done and effective methods developed to study light scattering by spheres. If a particle has a spherical symmetry, one can obtain the exact solution of the electromagnetic wave diffraction problem. If a particle is non-spherical, the problem is more complicated, and the number of publications on this subject is quite limited.[6,9]

When particle sizes are comparable with the incident wavelength, the problem becomes especially complicated, unless some simplifying assumptions are introduced. Generally, one of such assumptions is the small value of $(m-1)r$ as compared to the incident wavelength $\lambda$ ($r$ is the particle size, $m$ is the relative refractive index of a particle in a given dispersion). Another common assumption is the choice of a simple particle shape for description of light scattering.

A general formulation of the diffraction problem for a single particle is fairly simple.

The electric field $\boldsymbol{E}_0$ of a light wave excites the field $\boldsymbol{E}_\mathrm{i}$ inside a particle and the diffraction field $\boldsymbol{E}_\mathrm{s}$ outside it. The problem is to find the total field $\boldsymbol{E}$ using the Maxwell equations

$$\mathrm{rot}\boldsymbol{H} = \frac{4\pi\boldsymbol{j}}{c} + \frac{1}{c}\frac{\mathrm{d}\boldsymbol{D}}{\mathrm{d}t} \tag{1.1}$$

$$\boldsymbol{E} = -\frac{1}{c}\frac{\mathrm{d}\boldsymbol{H}}{\mathrm{d}t}, \tag{1.2}$$

where $\boldsymbol{D} = \epsilon\boldsymbol{E}$, $\boldsymbol{j} = K\boldsymbol{E}$, $\epsilon$ is the dielectric permittivity, $c$ is the vacuum velocity of the electromagnetic wave, $\boldsymbol{j}$ is the current density, $K$ is the electric conductivity, $\boldsymbol{H}$ is the magnetic field, $t$ is time. Inside the particle the total field $\boldsymbol{E}$ coincides with $\boldsymbol{E}_\mathrm{i}$, while outside the particle it equals the superposition of the fields $\boldsymbol{E}_0$ and $\boldsymbol{E}_\mathrm{s}$, i.e. $\boldsymbol{E} = \boldsymbol{E}_0 + \boldsymbol{E}_\mathrm{s}$.

Equations (1.1) and (1.2) should be supplemented by an expression that describes the charge conservation law

$$\mathrm{div}\boldsymbol{j} + \frac{\mathrm{d}\rho}{\mathrm{d}t} = 0, \tag{1.3}$$

where $\rho$ is the electric charge density. This expression can be conveniently re-written in the form

$$4\pi\mathrm{div}\boldsymbol{j} + \frac{\mathrm{d}}{\mathrm{d}t}\mathrm{div}\boldsymbol{D} = 0,$$

if one takes into account that

$$\mathrm{div}\boldsymbol{D} = 4\pi\rho. \tag{1.4}$$

Besides, to make the problem closed, we need the relation

$$\mathrm{div}\boldsymbol{H} = 0. \tag{1.5}$$

Here we assume the magnetic permeability of the material to be $\mu = 1$. In spite of its simple formulation, this problem is mathematically very complicated and can be solved only for a small number of cases.

A simple model of a scatterer in a three-dimensional space is a spherical homogeneous particle. A solution is possible, if the particles are electrically conductive, i.e. the refractive index $m$ may be complex. The set of equations (1.1), (1.2), (1.4), and (1.5) permits separation of variables for a sphere. This problem was analyzed by A. Low, J. Mie and R. Debye at the turn of the 20th century. The general theory of light scattering by spheres was termed as the Mie theory and has been described by several authors.[11,12] The solution was found analytically, but it has such a cumbersome form that a result can be obtained only using tables or complicated computations.

Since a spherical particle cannot be oriented by an electric field, the electrooptic problems for colloidal dispersions involve consideration of light scattering by non-spheres. A simple model of a non-spherical scatterer is a circular cylinder of an infinite length. The solution has been obtained for an arbitrary orientation of the cylinder axis relative to the electric and wave vectors of the electromagnetic wave. A detailed treatment of the problems for a homogeneous cylinder and an infinitely thin disk (another limiting case of a cylinder) is given in References 3,13,14. The separation of variables in differential equations for spheres and cylinders was accomplished using the spherical and cylindrical coordinates. However, this approach is no longer valid for such particles as biaxial and triaxial ellipsoids or for a cylinder of a finite length.[11,15] We mean the possibility of obtaining a complete analytical solution like the one derived from the Mie theory. Evidently, new approaches are necessary that would combine analytical and numerical methods for obtaining solutions for real non-spherical particles.

A possible approach to this problem is to use the collocation technique, in which the solution is represented as non-orthogonal series over special functions fitting a particular case. In this method, a sufficiently smooth surface is marked by a number of points, at which the boundary conditions must be satisfied. The number of terms in the series expansion is matched with the number of points on the surface in order to get a set of algebraic equations for the series coefficients.

Of special interest for the electrooptics of scattering media is the region of diffraction resonance, where the incident wavelength is comparable to the scatterer's size. For this case, it is very profitable to use the method of integral equations, in which the boundary value problems with elliptical equations are reduced to integral equations, using Green's theorem for vectors. One can ultimately derive the Fredholm equations of the first and second kind, which in turn can be solved numerically.[16] Integral equations are used to solve the diffraction problems for thin disks, spheroids, cones and truncated cones.

An alternative approach is the projection technique, in which the scatterer under study is placed in a sphere of radius $R$, and the field within the sphere volume $V$ is found using a series expansion over the basic functions of empty electric and magnetic resonators.[17] Outside the sphere, the scattered wave is represented as a series over a set of functions for diverging waves. Then the projection of Maxwell equations onto the space of the basic functions is considered, and the problem is reduced to a set of linear algebraic equations that are solved numerically. This technique gives a large error for large-angle scattered intensities but provides a fairly good value for the scattering coefficient.

## 1.2. LIGHT SCATTERING BY SMALL PARTICLES

The simplifying assumption on a small particle size allows one to consider a particle as being placed in a uniform electric field. This subject has been discussed at length by Van de Hulst.[3] The induced dipole moment $\boldsymbol{d}$ of a particle is defined as $\boldsymbol{d} = \hat{\gamma}\boldsymbol{E}_0$, where $\hat{\gamma}$ is the particle polarizability tensor. In the general case, the vector $\boldsymbol{E}$ is not parallel to $\boldsymbol{E}_0$, which is equivalent to polarization variation in the scattering process. The total energy scattered in all directions is

$$W = \frac{1}{3}k^4 c\,|\boldsymbol{d}|^2,$$

where $k$ is the wave number. The incident wave has the intensity

$$J = \frac{c}{8\pi}\,|\boldsymbol{E}_0|^2.$$

The scattering cross section $C_{\text{sc}}$ is defined as the intensity ratio of the scattered and incident radiations

$$C_{\text{sc}} = \frac{8}{3}\pi k^4 \overline{\gamma}^2,$$

where $\overline{\gamma}^2 = \ell_1^2\gamma_1^2 + \ell_2^2\gamma_2^2 + \ell_3^2\gamma_3^2$, $\ell_1$, $\ell_2$ and $\ell_3$ are the direction cosines of the vector $\boldsymbol{E}_0$ with respect to the principal axes of the tensor $\hat{\gamma}$; $\gamma_1$, $\gamma_2$ and $\gamma_3$ are the principal elements of $\hat{\gamma}$.

If the particles absorb light, the polarizability is a complex value, and for the absorption cross section defined by the absorbed-to-incident energy ratio, the following expression is valid

$$C_{\text{abs}} = 4\pi k \text{Re}(\text{i}\hat{\gamma}).$$

However, the majority of problems in the optics and electrooptics of scattering media requires knowledge of the total energy including both the scattered and absorbed energies. In this case one deals with the cross section of light extinction $C_{\text{ext}} = C_{\text{sc}} + C_{\text{abs}}$. Analysis of the cross sections $C_{\text{sc}}$ and $C_{\text{abs}}$ shows that if $\hat{\gamma}$ is independent of the wavelength, the order of these values is determined by the relations

$$C_{\text{sc}} \sim \lambda^4 v^2, \qquad C_{\text{abs}} \sim \lambda^{-1} v,$$

where $v$ is the particle volume. When $v \to 0$, the particles mainly absorb the light, whereas the share of the scattered energy increases with increasing $v$. For small particles with a low refractive index, to find $C_{sc}$ and $C_{abs}$ is a fairly easy problem.

The polarization vector $\boldsymbol{P}$ can be represented as

$$\boldsymbol{P} = \frac{m^2 - 1}{4\pi}\boldsymbol{E}. \tag{1.6}$$

Normally, the difficulty is that the vector $\boldsymbol{E}$ depends on the particle polarization, however, with small $(m^2 - 1)$, it can be replaced by $\boldsymbol{E}_0$. Then, by dividing equation (1.6) by $\boldsymbol{E}_0$ and by integrating it over the entire volume, we obtain the following expression for $\overline{\gamma}$:

$$\overline{\gamma} = \frac{1}{4\pi}\int_v (m^2 - 1)\mathrm{d}v.$$

For a homogeneous particle we can write

$$\overline{\gamma} = \frac{m^2 - 1}{4\pi}v.$$

It should be noted at this point that the substitution of $\boldsymbol{E}_0$ for $\boldsymbol{E}$ excludes the dependence of $C_{sc}$ on the $\boldsymbol{E}_0$ vector with respect to the principal axes $\hat{\gamma}$ in a system of oriented particles. In this case

$$C_{sc} = \frac{k^4 v^2}{6\pi}(m^2 - 1)^2$$

$$C_{abs} = -kv\left[\mathrm{Im}(m^2) - 1\right],$$

where $m^2 = \epsilon - 4\pi \mathrm{i} K'/\omega$, $\epsilon$ is the dielectric permittivity, $K'$ is the electrical conductivity of the particle material, $\omega$ is the circular light frequency. Thus, the solution of the scattering problem is reduced to the calculation of particle polarizability.

The simplest problem in the electrooptics of dispersions is that of light scattering by an ellipsoid. The field $\boldsymbol{E}_\mathrm{i}$ inside an ellipsoid can be represented by the expression $\boldsymbol{E}_\mathrm{i} = \boldsymbol{E}_0 + 4\pi L_j \boldsymbol{P}$, if the vector of $\boldsymbol{E}_0$ coincides with one of the ellipsoid axes and $(m^2 - 1)$ is not too small. Here, $L_j$ are the three factors defined by the semiaxis ratio and interrelated by the equality $L_1 + L_2 + L_3 = 1$. The expressions for $L_j$ at different semiaxis ratios can be found in Reference 3.

There are several approximations available in the light scattering theory for small particles with the refractive index $m$ close to unity. The Rayleigh–Debye–Gans (RDG) approximation is the one most commonly used, it is based on the assumption that

$$|m - 1| \ll 1, \qquad 2kr\,|m - 1| \ll 1.$$

In this approximation, it is possible to calculate the interference function of the field scattered by a particle with a given orientation, which is very important for electrooptic studies.

## 1.3. LIGHT SCATTERING BY LARGE PARTICLES

The approach to the solution of light scattering problems for large particles is qualitatively different from that for small particles. An incident light flux is considered as consisting of individual beams, each of which propagates along its own path independent of the other beams. According to Reference 3 the length $l_b$ of this path requires a beam width of about $\sqrt{l_b \lambda}$. For a beam to be considered as individual, its width must be larger than the light wavelength $\lambda$. This is possible when the particle sizes are by a factor of 20 larger than the incident wavelength. The light incident on such a particle is partly reflected and partly refracted, after which it is refracted again by another face of the particle and emerges from it. The light waves that have passed through a particle and those that have been reflected produce scattered light. It should be noted that there may be multiple internal reflections, but this fact does not alter the general formulation of the problem. Some of the incident energy is lost due to the light absorption by the particle. The amount of absorbed and scattered energy, as well as the light polarization and the scattering diagram vary with the properties of the material and its environment, with the shape, size and orientation of the particle with respect to the incident flux. Light beams passing in a close proximity of a particle experience Fraunhofer diffraction. The diffraction pattern also fully depends on the scatterer's shape, size and orientation in the scattering field but is independent of its material. For this reason, light-absorbing and light-reflecting bodies produce identical diffraction patterns. If the incident light is polarized, the polarization does not change following the diffraction. For a wavelength much smaller than the scatterer's size, the scattering diagram obeys the laws of geometrical optics, and the diffraction pattern has the shape of a narrow intense lobe extended along the light propagation direction. For larger wavelengths, the light intensity in the scattering diagram increases and becomes comparable to the diffraction pattern intensity at a given observation angle. In this case the theory of scattering by large particles fails to describe the optical phenomena if the refractive index $m$ differs considerably from unity. At $m$ close to unity, there is a certain range for the particle size-to-wavelength ratio, when the scattered intensity is comparable to the diffracted light and interferes with it. The total amplitude of the transmitted wave is defined by the sum of all the scattered wave amplitudes, because the phase difference between the electric field oscillations in the light fluxes is very small.

If a particle is sufficiently large compared to the wavelength, the light extinction cross section can be determined from Babinet's principle.[3] This principle, following from Huygens' principle, states that an opaque disk placed in the way of propagating light and a hole of the same shape in an opaque screen create perturbations in the incident light wave equal in magnitude and opposite in sign. It follows from this that

$$C_{\text{abs}} = 2G, \tag{1.7}$$

where $G$ is the geometrical cross section of the scatterer. This formula is valid only if the particle is sufficiently far from the photodetector and does not cast a shadow on it. Otherwise, one can resort to geometrical optics, from which $C_{\text{ext}} = G$.

Diffraction phenomena for large spheres and cylinders have been considered by M. Kerker.[5] However, the case we have just discussed is of little interest for electrooptic studies of dispersions, because particles with a radius $r > 10\lambda$ will sediment rapidly if the dispersion viscosity is low. Besides, there are more informative and simple ways of observing such particles with a microscope.

Of greater interest is scattering by particles with a low refractive index. These are largely biological dispersions, whose particles contain much water, so their $m$ is close to unity. Such particles have been called soft, in contrast to hard ones, for which $m$ differs much from unity.

Light scattering by particles in a situation when the inequalities $kr \gg 1$, $|m - 1| \ll 1$ are valid, have been considered by Van de Hulst. It has been shown that the interference and diffraction of the scattered light should be also taken into account. Van de Hulst's approximation, one advantage of which is its applicability to non-spherical particles, has been used to study light scattering by ellipsoids.[6,18]

## 1.4. LIGHT SCATTERING ANISOTROPY AND ELECTROOPTIC EFFECTS

Electrooptic effects are caused by variation in the optical properties of dispersions due to an electric field. These effects are a manifestation of dispersion anisotropy associated with particle orientation in an applied electric field. A system containing non-spherical oriented particles is very likely to exhibit other anisotropic features, for example, electrical conductivity or viscosity. Since the anisotropy of optical properties is best observed at low particle concentrations, dilute systems are very convenient model objects for electrooptic studies. We should note here that optical anisotropy which arises from interactions of particles, even spherical ones, in an electric field and the field-induced strains must also be identified as electrooptic effects.

Let us consider electrooptic effects in some detail. Dichroism of dispersions generally manifests itself when light absorption inside a particle is anisotropic and the directions of maximum and minimum absorption correlate in a specific way with the particle orientation axes in the external electric field. Dichroism is the only form of optical anisotropy that can be observed either in very small or in very large particles. In the latter case, the orientation of the polarization plane does not affect the light extinction coefficient, no matter how the particles are oriented.

Anisotropy of light scattering is associated with the correlation between the polarization plane orientation relative to the orienting field and the scattering cross section. This phenomenon covers a large class of non-spherical particles

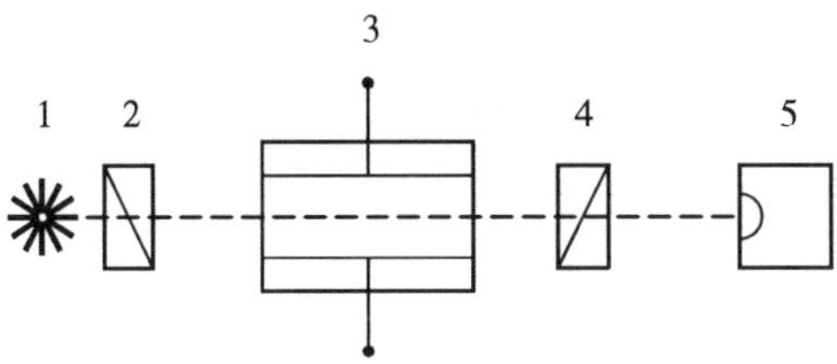


**Figure 1.1.** Schematic diagram of the experimental setup for birefringence studies.

varying in size from a few hundredths of a micrometer to 3–5 $\mu$m. It is especially distinct in the range of sizes close to the incident wavelengths; the range is generally termed as the diffraction resonance region.

Dichroism associated with scattering anisotropy and intrinsic dichroism are hard to separate experimentally, because both are related to variation in the intensity of light transmitted through a colloidal system, when an electric field is applied to it. This variation may be as large as 20–25% for particles of radius 0.5 $\mu$m and with a high refractive index, for example, in diamond and quartz dispersions with the material content of $10^{-2}$ mg cm$^{-3}$.

Light scattering in an external electric field is associated with variation in the scattered energy propagating in a given direction. A photodetector registers more radical changes in the light signal than in case of scattering anisotropy. However, the scattered intensity will strongly depend on the particle shape and the scattering angle, and the problem of finding the particle geometry from observations of electrooptic phenomena will have an unstable solution. The value of optical anisotropy is a function of the scattering cross section, which is defined by the intensity ratio of the incident and scattered light waves. The cross section is an integral characteristic which changes more smoothly than the scattering diagram. For this reason, in systems containing particles of variable shape, it is preferable to make observations of scattering anisotropy than of angular scattering.

Birefringence is connected with the correlation between the refractive index of a dispersion and the orientation of the light polarization plane with respect to the orienting field. The system behaves as a uniaxial crystal, whose optic axis is parallel to the applied field. In this case there is a phase difference between the light waves polarized parallel and perpendicular to the field. This produces elliptically polarized light at the outlet of the cell containing a dispersion, while at the inlet the light is polarized linearly. On the other hand, a phase difference may be detected in scattered beams if their polarizations are different, and the appearance of elliptical polarization may be due to the scattering specificity. Since there is no solution of the scattering problem for an arbitrarily shaped particle, it is impossible to separate birefringence from light scattering by strong scatterers.

Figure 1.1 shows a schematic diagram generally used for observation of birefringence. It consists of a light source (1), two polarizers (2 and 4), a cell (3) between them, and a photodetector (5). The axes of the polarizers are normal to each other and form an angle of 45° with the field induced by two parallel electrode plates placed in the cell. When the system in the cell is not optically anisotropic, because no field is applied to it, the light does not fall onto the detector. Generally, the appearance of birefringence is believed to be associated with the light falling on the detector, when a field is applied to the electrode plates. But if a dispersion possesses dichroism or scattering anisotropy, it behaves as a third (imperfect) polaroid placed between two crossed polarizers. This provides light incidence on the detector, no matter whether the dispersion has birefringence or not. Therefore, if a dispersion is primarily a light scatterer, observation of birefringence becomes very problematic. Possible errors in the study of birefringence of dispersions have been discussed by N. A. Tolstoi and P. P. Feofilov.[1]

Thus, we can conclude that the most commonly observed and correctly interpreted phenomena in colloidal dispersions are light scattering and anisotropy.

Development of a rigorous theory of electrooptic effects produced by light scattering in a dispersion is an extremely complicated and, for some cases, an unresolvable task. This is due to the fact that real dispersions are mostly polydisperse and contain particles of irregular shape. For scattering anisotropy, no quantitative theory is available at present, and one usually has to employ a phenomenological approach for a quantitative description of experimentally observed effects. S. P. Stoylov has summarized available data on the theory of light scattering by rods and disks oriented by an electric field.[19] However, the resultant formulae are too cumbersome and inconvenient for practical use. Therefore, it is reasonable to define relative changes in the physical quantity corresponding to an electrooptic effect, taking its maximum for fully oriented particles to be unity. In order to find a correlation between the relative variation in the electrooptic effect and the particle orientation, we must know how this effect changes when particles rotate relative to the orienting field.

Experimental tests with scattering and absorbing particles of irregular shape have shown that scattering-induced electrooptic phenomena are described by the relation

$$\frac{\Delta A}{A_0} \sim \frac{3\cos^2\theta - 1}{2}. \tag{1.8}$$

Here $A_0$ and $\Delta A$ are the light intensities scattered by fully and partially oriented particles, respectively; $\theta$ is the angle between the particle axis and the field. For completely oriented particles $\theta = 0$.

Expression (1.8) fairly well describes the experimental data, if the particle sizes do not exceed several wavelengths. It is valid for birefringence and dichroism, and in most cases does not yield considerable errors. Its applicability for a quantitative description of electrooptic effects has also been pointed out.[20]

Let us consider the behavior of a dispersion in a cell with two plate electrodes, to which an electric field is applied for some time. The field may

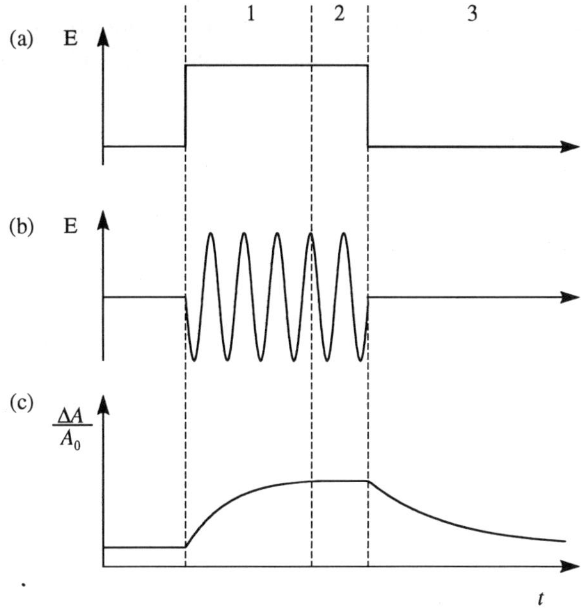


**Figure 1.2.** Time variation of the applied electric field and the recorded optical signal.

be constant, pulsed with sharp fronts or alternating (Figure 1.2a,b). With the applied field, the electrooptic effect (Figure 1.2c) starts to grow. Then, if the field is applied long enough, the system becomes steady-state, which corresponds to a time-independent electrooptic effect. After the field is turned off, the system relaxes into a state of random orientation, which corresponds to the time decay of the effect. These three stages are marked as 1, 2 and 3 in Figure 1.2. Region 2, the simplest one to interpret, corresponds to the time-independent angular distribution function $\mathcal{W}(\theta)$ of particle axial orientation. If the medium is electrically non-conductive, the function $\mathcal{W}(\theta)$ is described by the Boltzmann distribution law:

$$\mathcal{W}(\theta) = \frac{\exp\left[-u(\theta)/kT\right]}{\int_0^{\pi} \exp\left[-u(\theta)/kT\right] \sin\theta d\theta}. \tag{1.9}$$

Here $u(\theta)$ is the potential energy of a particle in the electric field. Its power expansion over the field strength, gives two terms:

$$u(\theta) = -\mu E \cos\theta - \frac{\gamma}{2} E^2 \cos^2\theta. \tag{1.10}$$

The coefficient $\mu$ is the permanent dipole moment and $\gamma$ is the anisotropy of the particle polarizability. The magnitude $\gamma$ is defined by the maximum variation of polarizability upon the particle rotation relative to the field $E$.

In electrically conductive dispersions, an applied orienting field induces current through the system, so the system will not be in equilibrium, though it may remain steady-state. The orientational order then is a function not only of the particle potential energy but of the flowing current as well. In this case we have no sufficient grounds to use the Boltzmann law to calculate the angular distribution of particles. The fact that the function $\mathcal{W}(\theta)$ cannot be expressed through $u(\theta)$ only is supported by the argument that the moment of orienting force cannot be represented as the potential energy derivative with respect to the angle $\theta$ because of the non-conservative nature of the conductive system, in which some of the electric energy is converted to thermal energy. The electric fields and hydrodynamic flows around a particle determine its angular rotation rate $\Omega$, which can be derived from equations for hydrodynamic flow rates and potential distribution. For particle rotation, the function of angular distribution of axes $\mathcal{W}(\theta)$ must be related to $\Omega$. This problem has been considered in detail by V. N. Shilov.[20] The function $\mathcal{W}(\theta)$ can then be described by the expression:

$$\mathcal{W}(\theta) \sim \exp \int_{\theta_0}^{\theta} \frac{\Omega(\theta)}{D} \mathrm{d}\theta, \tag{1.11}$$

where $D$ is the rotational diffusion coefficient of the particle.

In a non-conductive medium, where the orientation is related only to the dipole moment, the function $u(\theta)$ is defined by the expression

$$\int_{\theta_0}^{\theta} \frac{\Omega(\theta)}{D} \mathrm{d}\theta.$$

Thus, the dependence (1.11) is more general than (1.9) and is applicable to both conductive and non-conductive media. Nevertheless, we can introduce a more general concept of polarizability anisotropy, using the formula:

$$M = \gamma E_{\|} E_{\perp},$$

where $M$ is the total moment of forces acting on the particle, $E_{\|}$ is the field projection parallel to the particle orientation, and $E_{\perp}$ is the field projection normal to it. In this case, if $\mu = 0$, expression (1.9) can be still used for $\mathcal{W}(\theta)$, if $u(\theta)$ is given by (1.10). For a steady-state electrooptic effect, the dependence of $\Delta A/A_0$ on $\theta$ must be multiplied by $\mathcal{W}(\theta)$ and integrated over all $\theta$ directions. Due to the axial symmetry of $\mathcal{W}(\theta)$ and its dependence only on $\theta$, we finally have:

$$\frac{\Delta A}{A_0} = \int_0^{\pi} \frac{\Delta A(\theta)}{A_0} \mathcal{W}(\theta) \sin\theta \mathrm{d}\theta \tag{1.12}$$

or, with (1.8) and (1.9),

$$\frac{\Delta A}{A_0} = \frac{\Delta A_0}{2A_0} \frac{\int_0^{\pi} (3\cos^2\theta - 1) \exp\left[-u(\theta)/kT\right] \sin\theta \mathrm{d}\theta}{\int_0^{\pi} \exp\left[-u(\theta)/kT\right] \sin\theta \mathrm{d}\theta}. \tag{1.13}$$

Relaxation is a more complicated phenomenon for theoretical treatment. The time dependence of the electrooptic effect shown in Figure 1.2c (region 3) is largely determined by the geometry and optical characteristics of particles. But the effect corresponding to the steady-state orientation in Figure 1.2c (region 2) and described by (1.10) depends only on electrical characteristics of the particles. So one can get independent information about particles from the data on steady-state effects and on their relaxation processes.

In order to describe the relaxation behavior of $\Delta A/A_0$, the function $\mathcal{W}(\theta, t)$ should be introduced in (1.10) to take into account the relaxational variation in the particle orientation. This knowledge can be obtained from the diffusion equation for rotation:

$$\nabla^2 \mathcal{W}(\theta, t) = \frac{1}{D}\frac{\partial \mathcal{W}(\theta, t)}{\partial t}. \tag{1.14}$$

In the Laplace operator written in the spherical coordinates, only the terms dependent on $\theta$ should remain. Then the solution may take the form:

$$\mathcal{W}(\theta, t) = \sum_{n=1}^{\infty} C_n \exp\left[-n(n+1)Dt\right] P_n(\cos\theta). \tag{1.15}$$

Here $P_n$ are the Legendre polynomials over $\cos\theta$, $C_n$ are the coefficients defined by the initial condition of the steady-state orientation. By multiplying both sides of (1.15) by $P_n$ at $t = 0$ and by integrating over the range of cosine variation from minus unity to unity, we obtain the expression for $C_n$

$$C_n = \frac{2n+1}{n}\int_{-1}^{1} \mathcal{W}(\theta, 0) P_n(\cos\theta)\sin\theta d\theta.$$

There is only one summand remaining from the sum (1.15) due to the orthogonality of the Legendre polynomials. If expression (1.8) is valid, only the coefficient $C_2$ is non-zero; it is described by a formula analogous to (1.9) for a steady-state electrooptic effect. Then the general expression for $\Delta A/A_0$ takes the form:

$$\frac{\Delta A(t)}{A_0} = \frac{\Delta A_s}{A_0}\exp(-6Dt), \tag{1.16}$$

where $\Delta A_s/A_0$ corresponds to the steady-state effect in a given orienting field. If the particle sizes are so large that the $\Delta A/A_0$ ratio is proportional to $\cos\theta$ rather than to $P_2(\cos\theta)$, then (1.16) must be substituted by the expression:

$$\frac{\Delta A(t)}{A_0} = \frac{\Delta A_s}{A_0}\exp(-2Dt).$$

Region 1 in Figure 1.2 is the most difficult one to describe because of the simultaneous influence of the particle geometry and electrical properties on

$\Delta A/A_0$, while in regions 2 and 3 they behave independently. The expression for $\mathcal{W}(\theta, t)$ in this case can be obtained by solving the diffusion equation, in which we should include a term of force to describe the flow of particles being orientationally ordered by the field. Then we have:

$$\nabla^2 \mathcal{W}(\theta, t) + \frac{1}{kT} \mathrm{div}\,[\mathcal{W}(\theta, t)\mathrm{grad}u(\theta)] = \frac{1}{D}\frac{\partial \mathcal{W}(\theta, t)}{\partial t}.$$

The initial conditions at $t = 0$ correspond to a complete disorientation and at $t \to \infty$ are described by relation (1.9). The analytical solution of this equation for arbitrary values of $u(\theta)$ is unavailable. A solution in the form of a series can be found only for weakly oriented particles, when $u(\theta)/kT \ll 1$.

Growth of electrooptic effects due to light scattering in weak fields have been discussed.[19] We believe, however, that in most cases the consideration of effects in region 1 is practically unreasonable, because it is very hard to describe the region of growing effect and nearly impossible to get new information in addition to what we can derive from regions 2 and 3 in Figure 1.2.

A colloidal dispersion placed in a constant electric field is affected also by hydrodynamic orientation of particles due to their electrophoretic motion. For this case expressions (1.9), (1.10), and (1.13) become invalid. In order to avoid the electrophoretic effect on particle ordering, the alternating field frequency must be chosen in such a way that a particle would not be displaced at a distance comparable to its size for the time of field reversal. In such fields, the influence of the permanent dipole moment on the orientational order is also eliminated. Then the expression for $u(\theta)$ becomes simpler than (1.10):

$$u(\theta) = \frac{\gamma}{2} E^2 \cos^2\theta.$$

We will not further discuss the problems related to enhancement of electrooptic effects but focus on steady-state phenomena and their relaxation. These include effects produced by field rotation. Methods suitable for investigation of these problems have been described by several authors.[21,22] In this case the electric field vector rotates in a plane at the angular velocity $\Omega$. The particles follow it at the same angular velocity but with some phase delay. Some workers have thoroughly analyzed the particle motion in a rotating electric field.[23–25] They have studied steady-state rotation and transitional orientation modes produced by the field turn-on and turn-off. Orientationally ordered particles form a polarizer, although it is imperfect due to their dichroism. The rotation modulates the polarized light transmitted by the dispersion at the angular velocity $2\Omega$. Experimentally, the phase delay of particles from the field can be found from the Lissajous figures formed on the oscilloscope screen by the modulated optic signal and the sine-shaped voltage which induces the field in the cell.

# Chapter 2

# ELECTROOPTIC EFFECTS IN POLYDISPERSE SYSTEMS

## 2.1. STEADY-STATE ELECTROOPTIC EFFECTS

Steady-state electrooptic effects in electric fields are eventually described in much the same way, if the individual optical properties of particles are ignored. This Chapter is concerned with dichroism which, for most dispersions, reduces to scattering anisotropy, unless the particle material is dichroic.

It was pointed out in Chapter 1 that the amount of polarized light intensity scattered by non-spherical particles is a function of particle orientation. In colloidal dispersions the scattered light is largely incoherent,[5,26,27] and the dichroism problem is simplified because one may neglect interference phenomena.

Let us consider the transmission of linearly polarized light through a dispersion at various degrees of orientational ordering of irregular particles, i.e. for various spatial distributions of their axial orientation. We will do this in the Cartesian coordinate system. A linearly polarized light beam propagates along the $x$-axis and an external electric field $\boldsymbol{E}$ is along the $z$-axis (Figure 2.1). We assume the particle number density to be such that the dispersion satisfies Lambert–Beer's law and the light is polarized along the $z$-axis. Denote the extinction exponent for particles completely oriented along the $z$-, $y$-, and $x$-axes as $\mathcal{K}_1$, $\mathcal{K}_2$, and $\mathcal{K}_3$, respectively. For particles oriented along an arbitrary $l$-axis, whose direction is defined by the angles $\theta$ and $\varphi$, the extinction exponent can be found from the expression:

$$\mathcal{K}_{\|} = \mathcal{K}_1\cos^2\theta + \mathcal{K}_2\sin^2\theta\cos^2\varphi + \mathcal{K}_3\sin^2\theta\sin^2\varphi + S_{\|}(\theta,\varphi) + S(\theta,\varphi). \quad (2.1)$$

Subscript $\|$ shows that the light is polarized along the external electric field. Summands $S_{\|}(\theta,\varphi)$ and $S(\theta,\varphi)$ make up the fraction of $\mathcal{K}_{\|}$ which cannot be expressed as the products of $\mathcal{K}_1$, $\mathcal{K}_2$, and $\mathcal{K}_3$ and the squares of the respective direction cosines. The function $S(\theta,\varphi)$ is independent of the light polarization vector, unless particles possess intrinsic dichroism. If they do, the dependence of $S(\theta,\varphi)$ on the polarization vector can be included into the first three terms of expression (2.1). The function $S_{\|}(\theta,\varphi)$ is defined by the relative refractive index of the particle material in the medium and includes the interference components of coherent intensity scattered by the particle.

In case of incomplete orientational ordering, the extinction coefficient may be found, similarly to (1.12), from the expression

$$\mathcal{K}_{\|} = \int_{\omega} \mathcal{K}_{\|}(\theta,\varphi)\mathcal{W}(\theta)\mathrm{d}\omega, \quad (2.2)$$

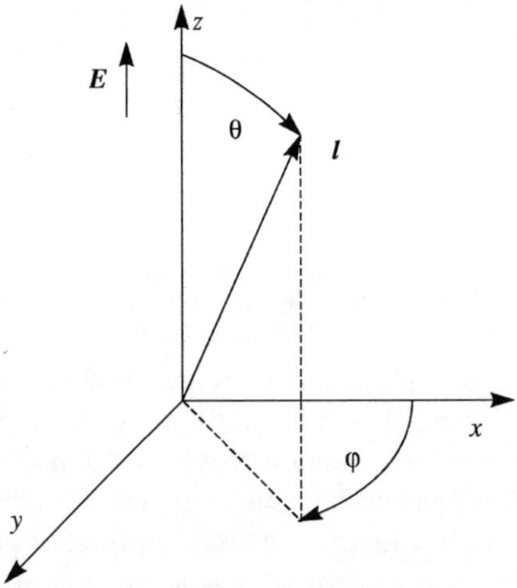


**Figure 2.1.** The orientation of the particle axis $l$ in the laboratory frame of reference.

where $d\omega = \sin\theta d\theta d\varphi$ is a component of the solid angle.

Since the field is along the $z$-axis, the distribution function $\mathcal{W}(\theta)$ has an axial symmetry and is independent of $\varphi$ (see Chapter 1).

By taking account of (2.1), (2.2) and averaging $\mathcal{K}_{\parallel}$ over $\varphi$, we obtain:

$$\mathcal{K}_{\parallel} = \int_0^{\pi} \left( \mathcal{K}_1 \cos^2\theta + \frac{\mathcal{K}_2 + \mathcal{K}_3}{2} \sin^2\theta \right) \mathcal{W}(\theta) \sin\theta d\theta + S_{\parallel} + S, \tag{2.3}$$

where

$$S_{\parallel} = \frac{1}{2\pi} \int_{\omega} S_{\parallel}(\theta, \varphi) \mathcal{W}(\theta) d\omega$$

$$S = \frac{1}{2\pi} \int_{\omega} S(\theta, \varphi) \mathcal{W}(\theta) d\omega.$$

However, the extinction coefficient is, by definition, equal to

$$\mathcal{K}_{\parallel} = \ln \frac{J_0}{J_{\parallel}}, \tag{2.4}$$

where $J_0$ and $J_{\parallel}$ are the intensities of the incident and transmitted polarized light, respectively. By substituting (2.4) into (2.3), we obtain:

$$\ln \frac{J_0}{J_{\parallel}} = \int_0^{\pi} \left( \mathcal{K}_1 \cos^2\theta + \frac{\mathcal{K}_2 + \mathcal{K}_3}{2} \sin^2\theta \right) \mathcal{W}(\theta) \sin\theta d\theta + S_{\parallel} + S. \tag{2.5}$$

For light polarization orthogonal to the electric field we may write, by analogy with (2.5):

$$\mathcal{K}_{\perp} = \ln \frac{J_0}{J_{\perp}} - \int_0^{\pi} \left( \mathcal{K}_2 \cos^2\theta + \frac{\mathcal{K}_1 + \mathcal{K}_3}{2} \sin^2\theta \right) \times \mathcal{W}(\theta) \sin\theta d\theta + S_{\perp} + S, \tag{2.6}$$

where $J_\perp$ is the intensity of light transmitted through the dispersion and polarized orthogonal to the field, and $S_\perp$ has the same meaning as $S_\parallel$ but only for normal polarization with respect to the electric field. It should be clear that $S_\parallel(\theta,\varphi)$ and $S_\perp(\theta,\varphi)$ may differ significantly, while $S_\parallel$ and $S_\perp$ do not differ so much. For dispersions with an isotropic distribution of particle axes over the angles $\theta$ and $\varphi$, $S_\parallel - S_\perp = 0$, but as the particles become ordered along the field, the difference $S_\parallel - S_\perp$ tends to

$$\frac{1}{2\pi}\int_0^{2\pi}\left[S_\parallel(\theta,\varphi) - S_\perp(\theta,\varphi)\right]\mathrm{d}\varphi.$$

The difference between $S_\parallel(\theta,\varphi)$ and $S_\perp(\theta,\varphi)$ depends only on the difference between the scattered coherent intensities for the parallel and perpendicular orientations of the polarization plane of incident light relative to the external field.

We further assume the quality

$$S_\parallel = S_\perp \tag{2.7}$$

to hold valid with an accuracy sufficient for correct analysis.

Equation (2.7) can be verified only experimentally, because there is no available theory of polarized light scattering for particles of arbitrary shape and of sizes comparable to the incident wavelength.

With (2.5), (2.6), and (2.7) we obtain

$$\ln\frac{J_\perp}{J_\parallel} = (\mathcal{K}_1 - \mathcal{K}_2)\int_0^{\pi}\frac{3\cos^2\theta - 1}{2}\mathcal{W}(\theta)\sin\theta\,\mathrm{d}\theta. \tag{2.8}$$

Expression (2.8) coincides with (1.13) with the accuracy to a constant factor. (Earlier, (1.13) was used to describe electrooptic effects due to birefringence and light scattering by particles with sizes much smaller than the incident light wavelength.)[20] In order to describe the electrooptic effect associated with dichroism induced by an electric field, it is expedient to introduce the magnitude

$$N = \frac{\ln(J_\perp/J_\parallel)}{\ln(J_\perp^0/J_\parallel^0)},$$

where $J_\perp^0$ and $J_\parallel^0$ are the values of $J_\parallel$ and $J_\perp$, respectively, for saturated particle ordering. It is evident that the value $N$ is suitable for description of the electrooptic effect, because it is independent of particle concentration in the dispersion or of the size of the cell. This makes the results from different experimental conditions comparable. It should be stressed that expression (2.8) was derived without restrictions imposed by particle sizes but with (2.7) being fulfilled. When the sizes exceed the incident wavelengths and the particles behave as strong scatterers, relation (1.13) still holds valid. The orientation of the polarization plane does not affect the extinction coefficient and $N$ is close to zero. However, it has been shown experimentally that the interference terms $S_{\|}(\theta,\varphi)$, $S_{\perp}(\theta,\varphi)$, and $S(\theta,\varphi)$ do not contribute much to extinction exponent so we can write:

$$\ln\frac{J_0}{J} = (\mathcal{K} - \bar{\mathcal{K}}) \int_0^{\pi} \frac{3\cos^2\theta - 1}{2} \mathcal{W}(\theta)\sin\theta d\theta,$$

where $\bar{\mathcal{K}}$ is the extinction coefficient for a disoriented system, and $\mathcal{K}$ is the same factor for a completely oriented system; $J$ coincides with both $J_{\|}$ and $J_{\perp}$.

In order to find the field dependence of an electrooptic effect for a steady-state case, one can make use of Boltzmann distribution for $\mathcal{W}(\theta)$. With expressions (1.9) and (1.10) for $\mathcal{W}(\theta)$ and $u(\theta)$, respectively, we can write:

$$\ln\frac{J_{\perp}}{J_{\|}} = \Delta\mathcal{K}\mathcal{F}(\chi,\lambda), \tag{2.9}$$

where

$$\Delta\mathcal{K} = \mathcal{K}_1 - \mathcal{K}_2$$

$$\mathcal{F}(\chi,\lambda) = \frac{3}{2}\frac{\int_0^1 x^2\exp\left(\chi x^2 + \lambda x\right) \mathrm{d}x}{\int_0^1 \exp\left(\chi x^2 + \lambda x\right) \mathrm{d}x} - \frac{1}{2} \tag{2.10}$$

$$x = \cos\theta, \quad \chi = \frac{\gamma E^2}{2kT}, \quad \lambda = \frac{\mu E}{kT}. \tag{2.11}$$

The calculation of $\mathcal{F}(\chi,\lambda)$ is a hard task at arbitrary $\chi$ and $\lambda$, except for very small values. The matter is that in direct calculations of (2.10) we deal with large number ratios close to unity, therefore, small errors in the evaluation of integrals from (2.10) may yield appreciable errors in $\mathcal{F}(\chi,\lambda)$. Application of this formula requires an algorithm convenient for computation of $\mathcal{F}(\chi,\lambda)$ with sufficient accuracy. Such algorithms, fitting small and large values of $u(\theta)/kT$, have been examined previously.[19] However, no computation of $\mathcal{F}(\chi,\lambda)$ with intermediate values has been made. Here we present an algorithm suitable for the computation of $\mathcal{F}(\chi,\lambda)$ at any $\chi$ and $\lambda$, which does not involve evaluation of integrals from (2.10).

Let us now consider three possible situations.

(i) The energy of particle interaction with the electric field is determined only by the induced dipole moment. This situation can be observed in studies of steady-state phenomena in sine-shaped high-frequency electric fields, when the period of oscillations is much shorter than the relaxation time of orientation due to the permanent dipole moment of particles and the relaxation time of field-induced anisotropy of the dispersion. If the field includes only a rapidly changing variable, no influence of $\mu$ on orientational ordering is observed, and $\lambda$ must be taken to be zero in expression (2.10).

The value of the electric field $E$ in (2.11) depends on both the magnitude and the mode of the applied field. For example, for a sine-shaped field, $E$ is the effective value.

Let us consider the calculation of $\mathcal{F}(\chi,\lambda)$ at $\lambda = 0$. We introduce the notations:

$$L(\chi) = \frac{I_2}{I_1}$$

$$I_1 = \int_0^1 \exp\left(\chi x^2\right) \mathrm{d}x, \qquad I_2 = \int_0^1 x^2 \exp\left(\chi x^2\right) \mathrm{d}x. \tag{2.12}$$

Then, with (2.10) and (2.12), we have:

$$\mathcal{F}(\chi,0) = \frac{3}{2}L(\chi) - \frac{1}{2}. \tag{2.13}$$

The function $L(\chi)$ turns out to satisfy the differential equation:

$$\frac{\mathrm{d}L}{\mathrm{d}\chi} = -L^2 + \left(1 - \frac{3}{2\chi}\right) L + \frac{1}{2\chi}. \tag{2.14}$$

This follows from the validity of the relations:

$$\frac{\mathrm{d}I_1}{\mathrm{d}\chi} = I_2, \quad I_2 = \frac{e^{\chi} - I_1}{2\chi}, \quad L(\chi) = \frac{e^{\chi}}{2I_1\chi} - \frac{1}{2\chi}.$$

Expressions (2.13) and (2.14) yield a differential equation for $\mathcal{F}(\chi,0)$:

$$\frac{\mathrm{d}\mathcal{F}}{\mathrm{d}\chi} = -\frac{2}{3}\mathcal{F}^2 + \left(\frac{1}{3} - \frac{3}{2\chi}\right)\mathcal{F} + \frac{1}{3}. \tag{2.15}$$

If at some $\chi_0 > 0$, the value of

$$\mathcal{F}_0 = \mathcal{F}(\chi_0,0) \tag{2.16}$$

is known, we can obtain the values of $\mathcal{F}(\chi,0)$ at any $\chi > \chi_0$ using, for example, the Runge–Kutta method numerical solution of equation (2.15).

Expansion of $\mathcal{F}(\chi,0)$ in power series at small $\chi$ provides $\mathcal{F}_0$. We re-write equations (2.15) and (2.16) in a form generally accepted for the Brio–Buce equation:[28]

$$\chi\frac{\mathrm{d}\mathcal{F}}{\mathrm{d}\chi} + \frac{3}{2}\mathcal{F} = \frac{\chi}{3} + \frac{\chi}{3}\mathcal{F} - \frac{2}{3}\chi\mathcal{F}^2, \quad \mathcal{F}(\chi,0) \to 0 \quad \text{at } \chi \to 0. \tag{2.17}$$

Although the Cauchy problem (2.17) has special initial conditions, it can be solved unambiguously and become holomorphic at sufficiently small $\chi$, i.e. representable as a converging series:

$$\mathcal{F}(\chi,0) = \sum_{k=1}^{\infty} a_k \chi^k. \tag{2.18}$$

Substituting (2.18) into (2.17) and using the method of uncertain coefficients, we have $a_1 = 2/15$, $a_2 = 4/315$ and:

$$a_k = \frac{3}{2k+1}\left(\frac{a_{k-1}}{3} - \sum_{j=1}^{k-2} a_j a_{k-j-1}\right). \tag{2.19}$$

For large values of $\chi$ the function $\mathcal{F}(\chi,0)$ can be re-written as follows:

$$\mathcal{F}(\chi,0) \simeq \tilde{\mathcal{F}}(\chi,0) = 1 - \frac{3}{2\chi - 1}. \tag{2.20}$$

At $\chi \geq 50$:

$$\frac{\left|\mathcal{F}(\chi,0) - \tilde{\mathcal{F}}(\chi,0)\right|}{\mathcal{F}(\chi,0)} < 10^{-4}.$$

With formulas (2.18), (2.15) and (2.20), we can make a simple program to compute $\mathcal{F}(\chi,0)$ with any accuracy.

(ii) The energy of particle interaction with the electric field depends only on the permanent dipole moment. This situation may arise when electrooptic effects take place in fields of rectangular pulses of sufficient duration, i.e. when the pulse duration is much longer than the time of the orientation relaxation due to the permanent dipole moment. In this case the permanent moment is the dominant factor in the particle orientation by the field. If the induced moment does not affect appreciably the orientational ordering, then $\chi = 0$ and $\mathcal{F}(\chi,\lambda) = \mathcal{F}(0,\lambda)$, according to expression (2.10). The function $\mathcal{F}(0,\lambda)$ can be determined by direct integration of (2.10):

$$\mathcal{F}(0,\lambda) = 1 + \frac{3}{\lambda^2} - \frac{3\coth\lambda}{\lambda}. \tag{2.21}$$

Series expansion of (2.21) at small $\lambda$ gives:

$$\mathcal{F}(0,\lambda) = \frac{\lambda^2}{15} - \frac{2\lambda^4}{315} + \ldots \simeq \frac{\lambda}{15}\left(1 - \frac{2\lambda^2}{21}\right).$$

(iii) Let us now consider the case, when the energy of particle interaction with the electric field is governed by the permanent and induced dipole moments comparable in magnitude. This may occur in dispersions placed in fields of rectangular pulses, whose duration is much longer than the relaxation time of the particle orientation. The parameters $\chi$ and $\lambda$ are non-zero. In order to calculate $\mathcal{F}(\lambda,\gamma)$, we introduce the notations:

$$T_1 = \int_{-1}^{1} \exp\left(\chi x^2 + \lambda x\right) \mathrm{d}x, \quad T_2 = \int_{-1}^{1} x^2 \exp\left(\chi x^2 + \lambda x\right) \mathrm{d}x.$$

By differentiating $T_1(\chi, \lambda)$ over $\chi$ and integrating by parts, we obtain:

$$\frac{dT_1}{d\chi} = T_2$$

$$2\chi T_2 = \exp(\chi + \lambda) + \exp(\chi - \lambda) - \frac{\lambda}{2\chi}[\exp(\chi + \lambda) - \exp(\chi - \lambda)] + \left(\frac{\lambda^2}{2\chi} - 1\right) T_1.$$

More cumbersome transformations, but similar to the case (i), lead to the differential equation:

$$\begin{aligned} \frac{d\mathcal{F}(\chi, \lambda)}{d\chi} &= -\frac{2}{3}\mathcal{F}^2(\chi, \lambda) + \left(\frac{1}{3} - \frac{2}{\chi} + 2V_1(\chi, \lambda) + 2V_2(\chi, \lambda)\right) \\ &\quad \times \mathcal{F}(\chi, \lambda) + \frac{1}{3} - \frac{1}{\chi} + \frac{3(1 - \lambda^2/\chi)}{4\chi^2} \\ &\quad + V_1(\chi, \lambda) + 3V_2(\chi, \lambda)\left(1 - \frac{2}{\chi} + V_1(\chi, \lambda)\right), \end{aligned} \tag{2.22}$$

where

$$V_1 = \frac{\exp(\lambda) - \exp(-\lambda)}{2(\chi - \lambda)\exp(\lambda) + 2(\chi + \lambda)\exp(-\lambda)}, \quad V_2 \equiv \frac{1}{4\chi}\left(\frac{\lambda^2}{2\chi} - 1\right).$$

If $\chi \neq 0$, equation (2.22) has no singularity and may be solved numerically. The function $\mathcal{F}(\chi, \lambda)$ at small $\chi$ can be found from the Taylor formula:

$$\mathcal{F}(\chi, \lambda) \simeq \mathcal{F}(0, \lambda) + \alpha\chi,$$

where

$$\alpha = \frac{4}{3} + 18\frac{\lambda^2 + 2}{\lambda^4} - 6\frac{\lambda^2 + 6}{\lambda^3}\coth\lambda - \frac{1}{3}\mathcal{F}(0, \chi).$$

Expressions (2.8) and (2.9) describing the field dependence of electrooptic effect have been obtained on the assumption of (2.7), which establishes the equality of $S_{\parallel}$ and $S_{\perp}$. For this, it is quite sufficient if the "coherent term" in the expression for the scattered polarized light intensity does not vary with the polarization plane orientation. The validity of (2.8) can be tested if we find from (2.9) the field dependence of the electrooptic effect and compare it with the respective experimental dependence. In order to see that expression (2.10) is applicable to description of electrooptic phenomena in dispersions, it is necessary to check its validity for systems with dominant scatterers and dominant absorbers. (One should keep in mind that all particles in dispersions scatter or absorb light to some extent.)

Aqueous suspensions of graphite and synthetic diamond provide excellent model systems with dominant absorbers and dominant scatterers, respectively.

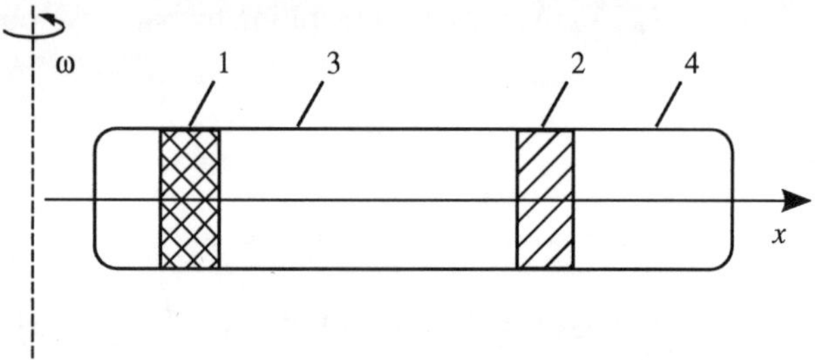


**Figure 2.2.** A test-tube for producing a monodisperse colloidal system.

Since expression (2.9) is valid only for monodisperse systems, the diamond and graphite suspensions in our experiments were made monodisperse by centrifugation. A test-tube with the suspension is shown in Figure 2.2. Sections 2, 3, and 4 of the test-tube, which was initially in the vertical position, were filled up with a pure dispersive medium. A standard polydisperse suspension with a high particle number density was added carefully, layer by layer, to section 1 without mixing it with the dispersive medium. Then the test-tube was placed in a centrifuge, in which it took the horizontal position, as shown in Figure 2.2, owing to the rotation at the angular velocity $\omega$. The volume of section 1 was made small enough for the coordinate $x$ of all particles to be approximately the same at the initial moment of centrifugation.

During the centrifugation, particles of various sizes moved along the coordinate $x$ at different velocities, because the forces acting on them varied with the particle size. Some time later, section 2 was found to be filled with particles of practically the same size. The size varied with the angular velocity $\omega$ and the centrifugation time $t$. From time to time, the centrifuge was stopped to take small samples of the diamond or graphite suspensions from section 2, using a syringe with a long needle. In this way we obtained a series of monodisperse samples with particles of about identical sizes, which was testified by an electron microscopic study. It should be stressed, however, that the procedure we used to obtain suspensions of a high degree of monodispersity is suitable only for large particles, since purely mechanical laws are applicable only to large particle motion in a centrifugal force field. The contribution of Brownian motion to the displacement of small particles should not be neglected.

A sine-shaped electric field was used to find the experimental dependences:

$$N = N(E^2). \tag{2.23}$$

The period of field oscillation ($2 \times 10^{-4}$ s) was much shorter than the relaxation time of the field-induced anisotropy due to thermal disordering of the particles ($10^{-2}$ s for a graphite suspension and $2 \times 10^{-2}$ s for a diamond suspension). The steady-state orientation here is due only to the induced dipole moment, whose value is defined by the anisotropy of particle polarizability and by the electric field.

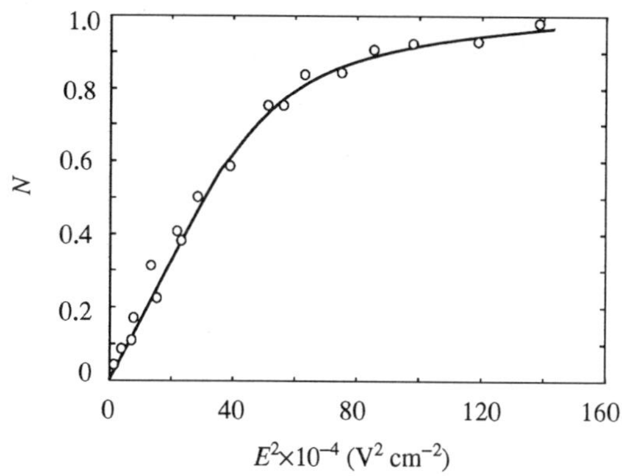


**Figure 2.3.** The function $N(E^2)$ for particles in an aqueous monodisperse diamond suspension.

All the measurements were made in white light from an incandescent lamp. The functions $S_{\|}$ and $S_{\perp}$ are integral values averaged over the wavelength in a wide spectral range. Since these functions depend only on the wavelength-to-size ratio, they are less sensitive to the size variation than in case of monochromatic light. For this reason, our study was restricted to white light and particle sizes of about 0.5 $\mu$m. The experiment showed that both suspensions possessed a distinct induced dichroism.

In the calculations of $N(E^2)$ from (2.9), one should take $\mathcal{F}(\chi, \lambda) = \mathcal{F}(\chi, 0)$, because the dipole moment $\mu$ does not affect the orientational order. Here and further, we will assume $N$, corresponding to complete ordering, to be equal to unity, when we compare the theoretical and experimental functions $N(E^2)$. Otherwise, no comparison would be possible, because one cannot calculate $\Delta\mathcal{K}$ in (2.8). A comparison of our calculations and experimental values has shown that expression (2.9) describes fairly well the experimental dependences for aqueous, practically monodisperse, suspensions of synthetic diamond and graphite. (In Figures 2.3 and 2.4a the theoretical functions $N(E^2)$ are shown by solid curves and the experimental data by dots.) This agreement supports the validity of the assumption in (2.7), which, however, does not imply that $S_{\|} + S$ and $S_{\perp} + S$ are always smaller than the first summands in (2.5) and (2.6). To illustrate the effect of these functions on the logarithmic extinction in the dispersions, Figure 2.4b presents the experimental curves for

$$\mathcal{K}_{\|} = \mathcal{K}_{\|}\left(E^2\right) \tag{2.24}$$

$$\mathcal{K}_{\perp} = \mathcal{K}_{\perp}\left(E^2\right) \tag{2.25}$$

for the graphite suspension. For the diamond suspension, the functions $\mathcal{K}_{\|}(E^2)$, $\mathcal{K}_{\perp}(E^2)$, and $N(E^2)$ are similar to these. It follows from Figure 2.4 that for some suspensions, e.g. for graphite, $N(E^2)$ should not be replaced by (2.24)

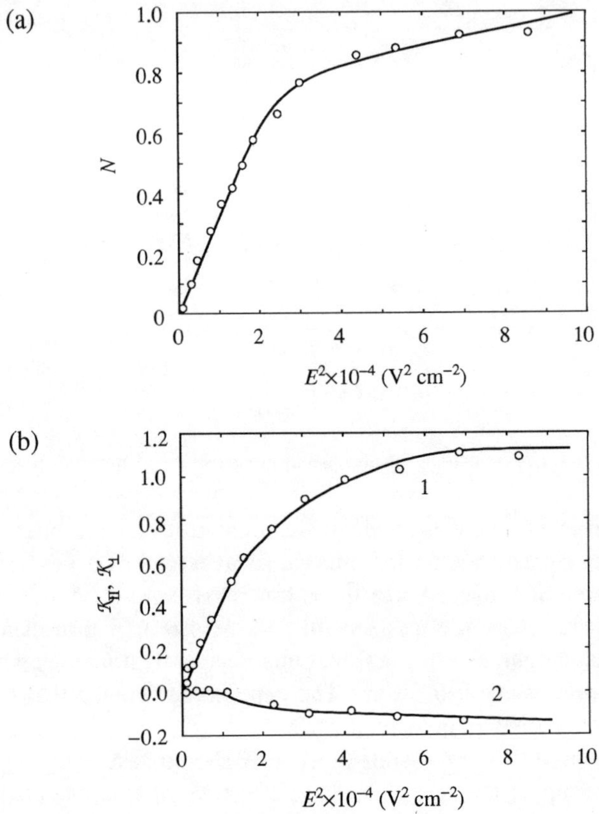


**Figure 2.4.** (a) The functions $N(E^2)$ for an aqueous monodisperse graphite system. (b) The functions $\mathcal{K}_{||}(E^2)$ (curve 1) and $\mathcal{K}_{\perp}(E^2)$ (curve 2) for the same system.

and (2.25) in the calculations of permanent moments and anisotropy of electrical polarizability.

From the comparison of the theoretical and experimental curves for $N(E^2)$ in Figures 2.3 and 2.4, one can find the value of $\gamma = 2\chi kT/E^2$. We have found the polarizability anisotropy for the above suspensions to be $1.9 \times 10^{-14}$ and $8.1 \times 10^{-12}$ cm$^3$ for diamond and graphite, respectively. In both cases the particle size was 0.5 $\mu$m.

## 2.2. THEORY OF INDUCED DICHROISM IN POLYDISPERSE SYSTEMS

Expression (2.9) is valid only for monodisperse colloidal suspensions, but to make dispersions with close particle sizes is a fairly complicated task. Perfect monodisperse systems have been obtained only with spherical particles. For this reason, expression (2.9) cannot be applied directly to most real dispersions and should be generalized for a polydisperse system.

If the polydispersity of a system is defined by the size distribution function $\varphi(r)$, an electrooptic effect is described by the expression:

$$N = \int_0^\infty \Delta\mathcal{K}\varphi(r)\mathcal{F}(\chi, \lambda)\mathrm{d}r. \tag{2.26}$$

The extinction coefficients of the polarized light transmitted by the dispersion, as well as the parameters $\chi$ and $\lambda$, are also the functions of the particle size, because both the permanent moment and the polarizability anisotropy are size-dependent. In order to evaluate $N(E^2)$ for a polydisperse system, one must know the dichroic-weighted size distribution function

$$f(r) = \Delta\mathcal{K}(r)\varphi(r) \tag{2.27}$$

and a particular form of the functions

$$\mu = \mu(r) \tag{2.28}$$

$$\gamma = \gamma(r). \tag{2.29}$$

A comparison of the experimental $N(E^2)$ values with the analytical ones can be used to find the particle size dependences of permanent moments and polarizability anisotropy. For this purpose, the function $N(E^2)$ should be found experimentally for a dispersion with a known distribution function $f(r)$. By assuming a particular form of the functions in (2.28) and (2.29) and by calculating $N(E^2)$ from (2.26), we can check the validity of the assumptions in these expressions from a comparison of the calculated and experimental values of $N$, and also derive the coefficients necessary to specify the electric properties of the particles.

It is clear that this approach is justified not only in determining the value of dichroism $N$, but it is also applicable to any study of particles with close sizes, if expression (1.13) is valid.

A correct solution of the problem of electrical characteristics of particles requires the solution of another problem: how to obtain the distribution function $f(r)$ in dichroism studies or its analog in studies of other electrooptic effects.

Let us consider the diffusion approach, since it is a universal and commonly used method for evaluation of the size distribution function $f(r)$. For the dichroic-weighted distribution function, this approach is based on analysis of the relaxation curve for optical anisotropy (region 3 in Figure 1.2). The curve shape corresponds to particle disorientation after the electric field is turned off. The rotational diffusion of particles, or their orientational relaxation, occurs regardless of the external action that produced the particle ordering. One should keep in mind that the process of orientational relaxation and, therefore, of the electrooptic effect relaxation become completely independent of the field only if all the particles were fully ordered at the moment of the field turn-off.

When the angular distribution of particle axes possesses an axial symmetry and the time dependence of this distribution is described by the function $\mathcal{W}(\theta, t)$, we can write, by analogy with (2.8), that

$$N(t) = \Delta\mathcal{K} \int_0^\pi \frac{3\cos^2\theta - 1}{2} \mathcal{W}(\theta, t) \sin\theta d\theta. \tag{2.30}$$

The normalization of the function $\mathcal{W}(\theta, t)$ is made with the equality

$$\int_0^\pi \mathcal{W}(\theta, t) \sin\theta d\theta \equiv 1.$$

Relation (2.30) may be re-written as follows:

$$N(t) = \Delta\mathcal{K} \left(1 - \frac{3}{2}\overline{\sin^2\theta}\right), \tag{2.31}$$

where

$$\overline{\sin^2\theta} = \int_0^\pi \mathcal{W}(\theta, t) \sin^2\theta \sin\theta d\theta.$$

If at the initial moment $t = 0$, the particles are completely oriented and $\mathcal{W}(\theta, t)$ represents a $\delta$-function, then $\overline{\sin^2\theta}$ varies with time according to the law:[29]

$$\overline{\sin^2\theta} = \frac{2}{3}[1 - \exp(-6Dt)], \tag{2.32}$$

where $D$ is the rotational diffusion coefficient of a particle. Simultaneous solution of (2.32) and (2.31) gives

$$N(t) = \Delta\mathcal{K} \exp(-6Dt).$$

This relation is valid only for monodisperse systems. Variation in the value of $N(t)$ in case of a polydisperse system does not obey the exponential law. Indeed, if the particle number density is low (the contribution of a single particle to the light flux modulation may be considered as independent), $N(t)$ is defined as follows:

$$N(t) = \int_0^\infty \Delta\mathcal{K} \exp[-6D(r)t] \varphi(r) dr, \tag{2.33}$$

where $\varphi(r)$ is the size distribution function. Taking into account the dichroic-weighted distribution function $f(r) = \Delta\mathcal{K}\varphi(r)$, expression (2.33) can be re-written in the form:

$$N(t) = \int_0^\infty f(r) \exp\left[-6D(r)t\right] dr. \tag{2.34}$$

If $f(r)$ is normalized to unity, then $N(t)$ varies from 1 to 0 with increasing $t$, and $N(0) = 1$. Using (2.34), one can find $f(r)$ if $D(r)$ is known and

$N(t)$ has been found experimentally. For this, one should solve the integral equation (2.34), which is the first-order Fredholm equation. Its kernel, $\exp[-6D(r)t]$, is a smoothly varying analytical function of the arguments $r$ and $t$. In the integration, the function $f(r)$ with the weight $\exp[-6D(r)t]$ is averaged, which makes $N(t)$ weakly dependent on the behavior of $f(r)$. In other words, the strongly differing functions $f(r)$ and $f_1(r)$ may correspond to the close functions $N(t)$ and $N_1(t)$, respectively. The specificity we have noted above is associated with the ill-posed problem in (2.34), whose solution is difficult but not impossible. Therefore, the mathematical method of solving equation (2.34) should be sufficiently sensitive to weak variations in $N(t)$.

Let us analyze the methods of solving this problem. If we replace the integral upper limit in (2.34) by a finite number $a$, for which the equality $f(r) = 0$ at $r \geq a$ is known beforehand, then equation (2.34) is transformed to:

$$N(t) = \int_0^a f(r) \exp\left[-6D(r)t\right] \mathrm{d}r. \tag{2.35}$$

Let us represent (2.35) as a set of linear equations:

$$\sum_{i=0}^{n} a_{ik} \exp\left[-6D(r_i)t_k\right] f(r_i)h = N(t_k), \quad k = 1, 2, \ldots, n, \tag{2.36}$$

where $r_i = ia/n$, $a_{ik}$ are the coefficients of the chosen formulas for numerical integration. By solving (2.36) we get the values of $f(r_i)$ at the points $r_i$. But the solution found in this way is strongly oscillating and does not correspond to the actual form of the function $f(r)$.

Equation (2.33) may be reduced to the Laplace transformation:

$$N(t) = \int_0^\infty \tilde{N}(p) \exp(-pt) \mathrm{d}p, \tag{2.37}$$

where

$$p = 6D(r), \qquad \tilde{N}(p) = \frac{1}{6} f(r) \left(\frac{\mathrm{d}D(r)}{\mathrm{d}r}\right)^{-1}.$$

Here we should assume that $\tilde{N}(+0) = 0$ and $\tilde{N}(p)$ exists at $\mathrm{Re}\,p > 0$. Then the problem (2.37) can be solved using the inverted Laplace transformation.

One of the possible ways of determining the original $\tilde{N}(p)$ is Papulis' method based on special expansions. Nevertheless, in most cases, as is shown in the work,[30] there is only a principal possibility to make the inverted Laplace transformation in series, but the ways to do this are not available. The sets of linear equations for deriving the coefficients of the series expansions are unstable.

An alternative way to derive the original $\tilde{N}(p)$ is to represent it in the form of the Mellin complex integral:

$$\tilde{N}(p) = \frac{1}{2\pi \mathrm{i}} \int_{c-\mathrm{i}\infty}^{c+\mathrm{i}\infty} N(t) \exp(pt) \mathrm{d}t, \tag{2.38}$$

where $c$ is the abscissa in the half-plane of the absolute convergence of the Laplace integral $t = c + i\tilde{\rho}$, $-\infty < \tilde{\rho} < +\infty$. The function $N(t)$ in (2.38) must be regular and analytical at Re$t > c$. Consequently, for experimental $N(t)$ one must find an approximation in the form of a half-plane regular function. The ways of evaluating the Mellin integral are explained explicitly in the work by Krylov.[30] The preferable ways are based on the use of formulas for numerical integration. The Mellin integral problem can be reduced also to the inverted Fourier transformation. However, numerical methods for solving equation (2.34) with further information on the solution and with account of the initial data error[31] are now considered as advantageous to integral transformations. In this case the problem in question may be referred to the class of conditionally correct ones.[32]

According to one of these methods,[31] equation (2.35) is substituted by a set of linear equations:

$$\sum_{i=0}^{n} a_{ik}\exp[-6D(r_i)t_k]hf(r_i) = N(t_k) + \epsilon_k,$$

$$k = 0, 1, \ldots, n,$$

where $\epsilon_k$ is the calculation error for $N(t_k)$.

Among the solutions of the class $F$, for which the function $N(t)$ is known with an error $\epsilon$, one chooses a solution of $f_s$ which is the smoothest and satisfies the condition

$$\int_0^a \left(\frac{d^2 f_s(r)}{dr^2}\right)^2 dr = \min\left\{\int_0^a \left(\frac{d^2 f}{dr^2}\right)^2 dr\right\}, \quad f \in F.$$

Another procedure, which takes into account *a priori* the features of $f(r)$ as additional information, is the regularization technique.[16,32] Comparison of the results obtained by these two methods for the model functions has shown that the regularization technique yields a more rigorous solution. We have chosen this technique in order to obtain the dichroic-weighted size distribution function.

Expression (2.35) is similar to (2.34), because one can always point out the size limit of particles in a given dispersion. We will seek for the solution in the Sobolev space $W_2^1$ with a metric defined by the formula:

$$\| f_1 - f_2 \| = \left\{\int_0^a \left[(f_1 - f_2)^2 + \left(\frac{df_1}{dr} - \frac{df_2}{dr}\right)^2\right] dr\right\}^{1/2}, \qquad (2.39)$$

where $f(r)$ is assumed to be a continuous smooth function; $f_1$ and $f_2$ are elements of space $W_2^1$, in which the solution is to be found. According to the theory,[32] the solution of $f(r)$ in this case may be reduced to determining a smoothing functional minimum

$$M^\alpha\left[f(r), N(t)\right] = \int_0^t \left( \int_0^a \exp[-6D(r)t] f(r) \mathrm{d}r - N(t) \right)^2 \mathrm{d}t + \alpha \int_0^a \left[ f^2(r) + \left( \frac{\mathrm{d}f(r)}{\mathrm{d}r} \right)^2 \right] \mathrm{d}r,$$

where $\alpha$ is the unknown parameter. A variational method can be used to define $f(r)$, which minimizes the functional $M^\alpha\left[f(r), N(t)\right]$. Taking the first variation of the functional $M^\alpha$ to be zero and considering the *a priori* data $f(0) = f(a) = 0$, we obtain an integral-differential equation

$$\int_0^a \hat{K}(r,\rho) f(\rho) \mathrm{d}\rho - \alpha \left( \frac{\mathrm{d}^2 f(r)}{\mathrm{d}r^2} - f(r) \right) = b(r), \qquad (2.40)$$

where

$$\hat{K}(r,\rho) = \frac{1 - \exp\{-6[D(r) + D(\rho)]t\}}{6[D(r) + D(\rho)]}$$

$$b(r) = \int_0^{t_0} N(t) \exp\left[-6D(r)t\right] \mathrm{d}t,$$

$t_0$ defines the interval $(0, t_0)$ in which the value of $N(t)$ is measured experimentally. Expression (2.40) belongs to the class of correct problems and its numerical solution can be found for any $\alpha$. For this purpose, equation (2.40) should be substituted by a set of linear equations. If we use a mesh with a step $h$, the set of equations takes the form:

$$\sum_{k=1}^{n-1} c_{ik} f_i = b_i \qquad (i = 1, 2, \ldots, n-1), \qquad (2.41)$$

where

$$f_i = f(ih), \qquad h = a/n$$

$$b_i = \int_0^{t_0} N(t) \exp\left[-6D(ih)t\right] \mathrm{d}t$$

$$c_{ik} = s_k h \frac{1 - \exp\{-6[D(ih) + D(kh)]t_0\}}{6[D(ih) + D(kh)]} + R_{ik}$$

$$R_{ik} = \begin{cases} \alpha/h^2 & \text{at } k^2 = i^2 - 1 \\ 2\alpha/h^2 + 1 & \text{at } k = i \\ 0 & \text{at other } k \end{cases}$$

Here $s_k$ are the coefficients of the numerical integration formulas. The regularization parameter $\alpha$ can be derived from the equation:[32]

$$\varphi(\alpha) = \left( \int_0^{t_0} \{\exp[-6D(r)t]\, f^{\alpha}(r) - N(t)\}^2 \,\mathrm{d}t \right)^{1/2} = \delta N, \qquad (2.42)$$

where $\delta N$ is an error described by the expression:

$$(\delta N)^2 = \int_0^{t_0} [N(t) - N_0(t)]^2 \,\mathrm{d}t.$$

Note that $N_0(t)$ is the exact function. Let us determine experimentally:

$$\epsilon_i^2 = [N(t_i) - N_0(t_i)]^2, \qquad (2.43)$$

where $t_i$ are the values of $t$, for which $N$ has been found. With (2.43) and (2.42), we can write down:

$$(\delta N)^2 = \sum_j \epsilon_j^2 \beta_j.$$

The coefficients $\beta_j$ depend on the interval between the values of $t$, for which $N$ is measured, and on the form of numerical integration formulas. After $N(t)$ and $\delta N$ have been obtained, we can determine $f(r)$ by simultaneously solving equations (2.41) and (2.42). For a certain sequence of $\alpha_1, \alpha_2, \ldots, \alpha_k$, we find the corresponding sequence of $f^{\alpha_1}(r), f^{\alpha_2}(r), \ldots, f^{\alpha_k}(r)$, from which we should select an element of space $W_2^1$, for which the equality

$$\varphi(\alpha) - \delta N = 0 \qquad (2.44)$$

is the closest one.

In order to test the results, we have used the regularization method to obtain the solution for a model function:

$$\tilde{f}(\tilde{r}) = \begin{cases} (\tilde{r} - \tilde{r}_0)\exp\left[-(\tilde{r} - \tilde{r}_0)^3\right] & \text{at } \tilde{r} \geq \tilde{r}_0 \\ 0 & \text{at } \tilde{r} < \tilde{r}_0 \end{cases}$$

was taken as a model with the parameter $\tilde{r}_0 = 0.08$. (Here $\tilde{f}$ and $\tilde{r}$ are dimensionless values.) This function is often used in analytical descriptions of size distribution of particles in dispersions. In order to derive $N(t)$ for the function $\tilde{f}(\tilde{r})$, it is sufficient to substitute (2.39) into (2.35) and evaluate the integral by means of the numerical integration formulas. We have calculated this dependence, assuming $\tilde{D}(r) = 1/\tilde{r}^3$ for simplicity. For real dispersions, $\tilde{D}(\tilde{r})$ differs from the assumed function only by a constant factor. The error in

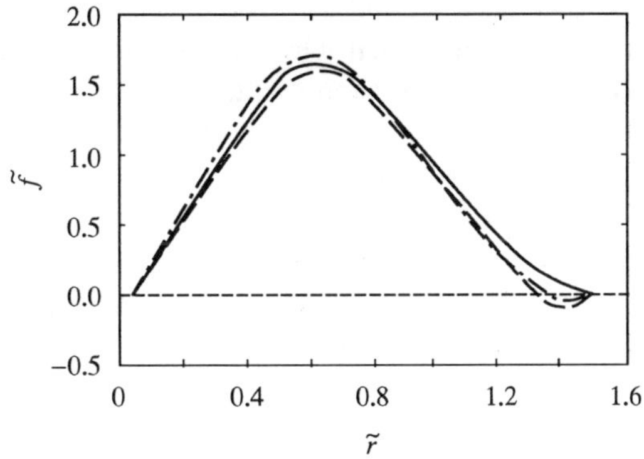


**Figure 2.5.** Illustration of the regularization technique: model function $\tilde{f}(\tilde{r})$ (solid curve), the solutions of an inverse problem for $N(t)$ error of 5% (dashed curve) and 10% (dash-dotted curve).

$N(t)$ was simulated with a special computational procedure to obtain random numbers with a normal distribution function.

Figure 2.5 presents the calculation results of $\tilde{f}(\tilde{r})$ obtained by the regularization technique for cases when $N(t)$ is known with a relative error of 5% (dashed curve) and 10% (dash-dotted curve). One can see that these functions are close to the exact function (2.27) shown by a solid line and that the difference between them decreases with decreasing error in $N(t)$. This illustrates the applicability of the regularization technique in calculations of a dichroic-weighted size distribution function, so we may expect to obtain reliable results from the study of real dispersions.

Sometimes it is expedient to find a solution among piece-wise smooth functions in the $C_1$ space and to replace (2.40) by the Fredholm integral equation of the second kind:

$$\int_0^a \hat{K}(r,\rho)f(\rho)\mathrm{d}\rho - \alpha\,[f(r)] = b(r),$$

where $\hat{K}(r,\rho)$ and $b(r)$ are the same functions as in (2.40). This equation can also be solved numerically by substituting it by a set of linear algebraic equations. In the calculation of $\alpha$, one can employ a procedure for finding the function minimum $\|\mathrm{d}f/\mathrm{d}\alpha\|_{C_1}$ as well as the residual method (see (2.42)).

To turn back to electrooptic effects, experiments on the polarization of colloidal particles and the electrical characteristics that determine it must include the measurement of $N$ as a function of time $t$ and electric field $E$. The polydispersity of colloidal systems in dichroism studies is characterized by the dichroic-weighted size distribution of particles. In a light scattering system with a particle size close to the incident wavelength, the scattering anisotropy is very pronounced, and the experimental data on its relaxation behavior can be quite accurate. The measurements of $N(t)$ are made in a setup which permits variation of the amplitude, duration and frequency of the electric pulses. In such studies, it is better to use a sine-shaped field with a frequency of several kilocycles per second in order to avoid a possible hydrodynamic change in the particle orientation due to the electrophoretic motion. It is necessary to achieve a complete ordering, so that formula (2.35) holds valid. Otherwise, particles of various sizes will have different degrees of orientational order, so the problem of finding $f(r)$ will become more complicated and less rigorous because of the necessity to know the size dependence of polarizability anisotropy. The pulse frequency must be low enough for the system to become completely disoriented over the period between two subsequent pulses. Higher pulse frequencies make the problem of finding $N(t)$ less accurate. The ordering of optically anisotropic particles at the moments of the field turn-on and their disordering at zero-field will produce modulation of the linearly polarized light transmitted by the dispersion. The modulation curve can be obtained with an oscilloscope and recorded for further analysis. The recording is made with a memory oscilloscope or a pulse time storage, which provides a higher accuracy of the modulation analysis. In order to find the relaxation curve of the electrooptic effect, the modulation curves of the polarized light transmitted by the dispersion are studied experimentally, starting with the termination of the orienting electric pulse. From the measured intensities $J_{\|}(t)$ and $J_{\perp}(t)$ of the transmitted light polarized parallel and perpendicular to the field during the particle disorientation, one can find $N(t)$ with a subsequent calculation of $f(r)$.

Figure 2.6 illustrates the experimental curves of $N(t)$ for aqueous colloidal dispersions of synthetic diamond and anisilidenbenzidine, respectively, and the curves of $f(r)$ calculated by the regularization technique for the same dispersions. The experimental error in $N(t)$ is a critical factor in determining the regularization parameter $\alpha$ by the residual method. It is associated primarily with the signal errors in $J_{\|}(t)$ and $J_{\perp}(t)$.

The constant components of optical signals can be measured with sufficient accuracy, with an error less than 0.1–0.2%. The absolute error in the alternating component is generally equal to that in the constant component. Hence, a relative error for the alternating components of an optical signal measured by a pulse accumulator is larger than for the constant components. It may be reduced down to 1–3% and even less by using an incandescent lamp as a light source. The light source is usually the principal contributor to the error in $N$. For a 2% error corresponding to the $N(t)$ curves in Figure 2.6, one can give the following equation:

$$\delta N = 0.02 \left( \int_0^t N^2(t) \mathrm{d}t \right)^{1/2} . \tag{2.45}$$

The value of $\delta N$ derived from (2.45) can be used to find from (2.44) the regularization parameter $\alpha$ necessary to calculate $f(r)$ for a given dispersion. If $f(r)$ is calculated at 40 to 50 points on the $r$ scale, $\alpha$ usually lies within the limits $10^{-2}$ to $10^{-4}$.

(a)

(b)

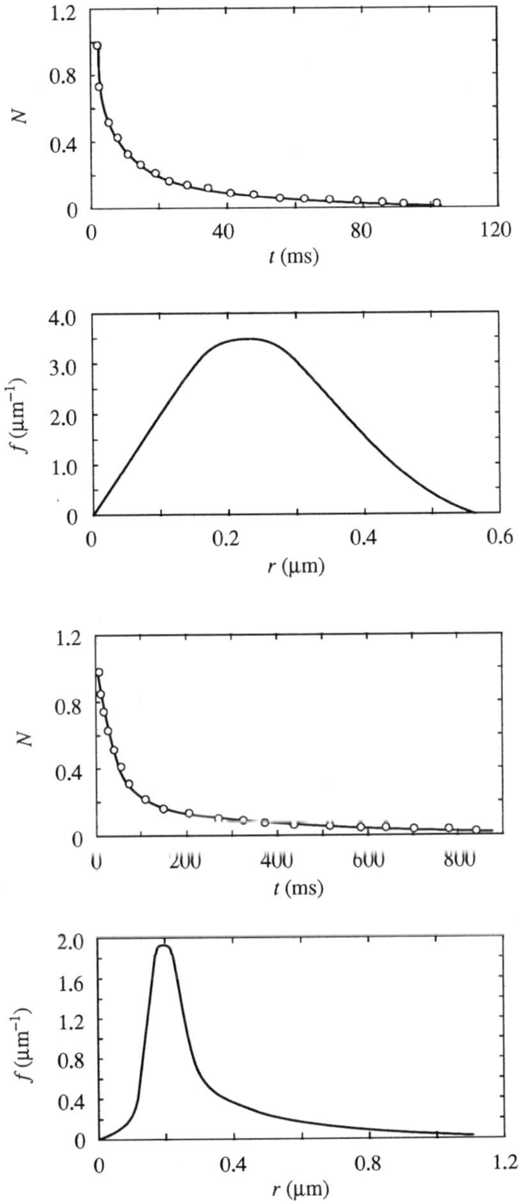


**Figure 2.6.** (a) The functions $N(t)$ and $f(r)$ for anisilidenbenzidine particles in water. (b) The same for diamond particles.

In calculating $f(r)$, it is necessary to pass from the rotational diffusion coefficient $D$ entering expression (2.35) to the $r$ scale, which requires knowledge of the particle shape. But when a particle rotates in a viscous medium, its shape

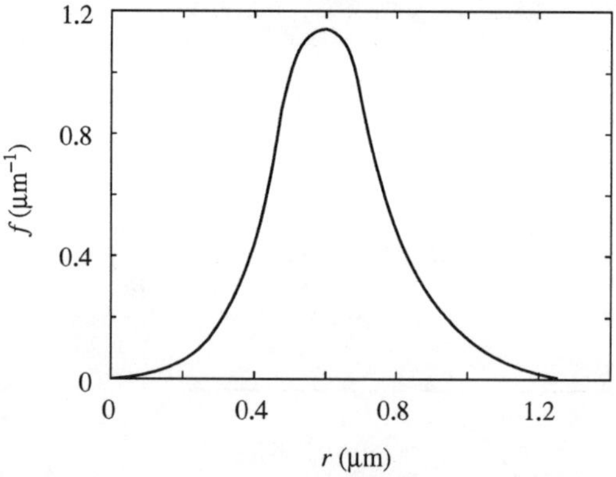


**Figure 2.7.** The function $f(r)$ for lunar soil aggregates in water.

does not affect the final result of the calculation as much as in case of light scattering characteristics. So in most cases the particles can be approximated by ellipsoids of revolution. The rotational diffusion coefficient then has the form:

$$D = \frac{kT}{[p]\eta V}.$$

The particle volume $V$ is calculated in the same way as for spheres of radius $r$, the structure factor $[p]$ can be found for spheroids with the semiaxis ratio $p$. The structure factor varies smoothly with the spheroid shape, and a permissible error in $p$ is found to be 20–30% , even if $p$ is close to unity.

The diffusion approach can be successfully used also to calculate $f(r)$ for stable particle aggregates. For example, Figure 2.7 illustrates the function $f(r)$ for particle aggregates in an aqueous suspension of rigolit (the lunar soil).

It should be noted that the diffusion technique can be applied not only to dichroism measurements, but also to angle-dependent scattered intensity and birefringence, or to any other magnitude that quantitatively describes an electrooptic effect and satisfies formula (1.8). In this case too the above approach to the description of electrical characteristics of polydisperse colloidal systems remains applicable.

## 2.3. ELECTROOPTIC STUDY OF POLYDISPERSITY

Information about the size distribution of particles is necessary for the solution of a wide variety of scientific and technical problems. Such problems regularly arise in commercial production of various materials, such as polishing pastes, coatings, synthetic diamond powders and other abrasives, ferrite powders for production of magnetics and magnetic coatings, colloidal cements, etc.

There are a few elaborate methods based on the study of birefringence and light scattering, which are commonly used to find the size distribution function of particles with a size much smaller than the incident wavelength.[33,34] When the sizes are comparable to the light wavelength, electron microscopic analysis and spectrophotometric techniques are employed.[33–35] However, electron microscopy does not permit measurements to be made directly in the experimental solution, without removing the particles. A serious disadvantage of this technique is that it damages the dispersion integrity, which results in a distorted picture of the size distribution of particles. Besides, only a small fraction of dispersed material can be analyzed in this case. For this reason, spectrophotometric techniques are more commonly employed involving small-angle light scattering,[26,36–39] depolarization of the scattered light[40] and nephelometric measurements.[41,42] All of these methods, however, are based on the theory of light scattering applicable only to spheres. They assume the size distribution function to be given *a priori* in an analytical form, so that the problem solution reduces to finding one or two parameters.[43] For this reason, nephelometric techniques provide in most cases only approximate data.

The development of electrooptic techniques to study colloidal dispersions and suspensions gives an opportunity of finding the size distribution function for non-spheres directly in a dispersion. Let us consider the calculation of polydispersity of a system from induced dichroism data at various $\lambda$, using the dichroic-weighted size distribution functions $f(r)$.

In the theory of scattering[3] by spheres, the extinction exponent $\mathcal{K}$ can be given in the form:

$$\mathcal{K} = \pi r^2 Q(x) n_0 \ell,$$

where $r$ is the particle size, $x = 2\pi r/\lambda$, $\ell$ is the light path in the cell. The extinction coefficient $Q(x)$ strongly depends on the refractive indices of the material and the dispersive medium and has a complicated non-monotonic character.

At present, no theory is available to describe polarized light scattering by particles of an arbitrary shape and comparable in size to the incident wavelength. However, one may assume that the change in the particle orientation with respect to the light vector is equivalent to a change in the particle's $Q$. In other words, one may suggest that

$$\mathcal{K}_1 = \pi r_0^2 Q_1(x_0) n_0 \ell \tag{2.46}$$

$$\mathcal{K}_2 = \pi r_0^2 Q_2(x_0) n_0 \ell, \tag{2.47}$$

where $x_0$ is calculated with the size $r_0$ equal to the radius of a sphere equivalent to the particle's volume (the particles are approximated by ellipsoids of revolution). From (2.46) and (2.47) we have:

$$\Delta\mathcal{K}(r_0) = \mathcal{K}_1 - \mathcal{K}_2 = \pi r_0^2 Q_1(x_0)\left(1 - \frac{Q_2(x_0)}{Q_1(x_0)}\right) n_0 \ell. \tag{2.48}$$

Relation (2.48) shows the dichroic weight to depend in a complex way on the particle size and the wavelength. However, for some cases this expression

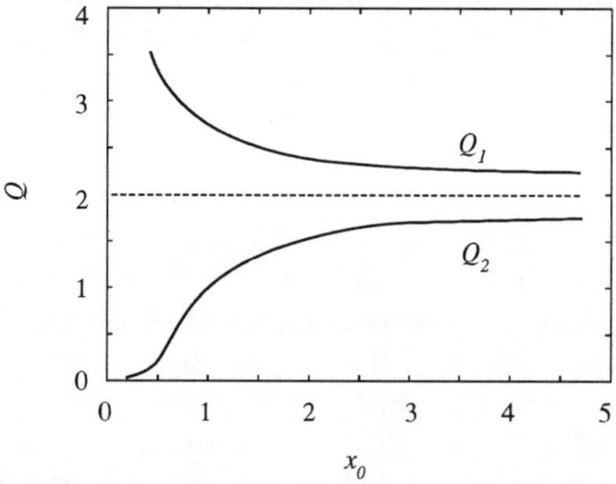


**Figure 2.8.** $Q_1$ and $Q_2$ versus $x_0$ for conductive cylindrical particles.

can be considerably simplified. Figure 2.8 presents the functions $Q(x_0)$ for electrically conductive particles and shows[3,44] that the extinction coefficients become practically independent of $x_0$, beginning from $x_0 \approx 1$. This means that for the visible light, $Q$ values are practically independent of the particle sizes larger than 0.1 $\mu$m. Therefore, electrooptic phenomena in dispersions with conductive particles, whose sizes satisfy the inequality $2\pi r_0/\lambda > 1$, are independent of the incident wavelength.

The study of electrooptic properties of an aqueous colloidal solution of graphite has supported these considerations. The dichroic-weighted size distribution functions $\Delta\mathcal{K}$ for graphite are practically independent of the incident wavelength $\lambda$ in the range from 402 to 723 nm (Figure 2.9). Hence, we may consider that for this colloidal dispersion $\Delta\mathcal{K} = c\pi r^2$, where $c$ is independent of $r$ and $\lambda$, and the particle shape, or the $a/b$ ratio, is assumed to be independent of the particle size. With account of (2.27), we obtain:

$$\varphi(r) = \frac{f(r)}{c\pi r^2}. \tag{2.49}$$

Expression (2.49) enables one to find the function $\varphi(r)$ with a known $f(r)$.

The curves $f(r)$ and $\varphi(r)$ for the system under study are presented in Figure 2.10a. Control electron microscopic observations of graphite particles have shown that $\varphi(r)$ well describes their size distribution.

According to Reference 3, which studied light scattering by long rod-shaped particles, the scattering by a single particle is proportional to its length $L$:

$$\Delta\mathcal{K} = AL, \tag{2.50}$$

where $A$ is the proportionality factor. If $\varphi(r)$ for a system is known, its form can be found using (2.50):

$$\varphi(L) = \frac{f(L)}{AL}.$$

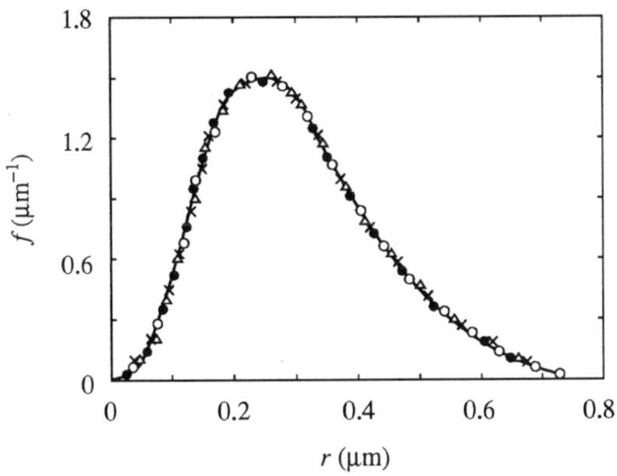


**Figure 2.9.** The function $f(r)$ for an aqueous graphite dispersion at various wavelengths. $\times$, 402 $\mu$m; $\triangle$, 458 $\mu$m; $\bigcirc$, 561 $\mu$m; $\bullet$, 723 $\mu$m.

We have obtained the functions $f(L)$ and $\varphi(L)$ for a palygorskit aqueous colloidal solution (Figure 2.10b). The function $\varphi(L)$ is in good agreement with electron microscopic data.

We will now turn to the case of "diffraction resonance", when the particles are comparable in size with the wavelength of the incident light, and the relative refractive index of the dispersive medium and the particle is $m \approx 2$. Figure 2.11 shows the function $Q(x)$ for this case to be complicated and non-monotonic. We re-write (2.48) in the form:

$$\Delta K = \pi r^2 F\left(\frac{r_0}{\lambda}\right), \tag{2.51}$$

where the function $F(r/\lambda)$ describes the dependence of the scattering cross section on the wavelength. One should note that at a given refractive index $m$, $F$ depends only on the ratio $r/\lambda$. A study of the relaxation of the electrooptic effect has yielded various values of $f(r)$. By giving a certain set of values to $\lambda_i$ ($i = 1, 2, \ldots n$), we get, according to (2.51), a set of functions:

$$f_i(r) = \pi F\left(\frac{r}{\lambda_i}\right) r^2 \varphi(r). \tag{2.52}$$

If $\varphi(r)$ is normalized by

$$\int_0^a \varphi(r) \mathrm{d}r = 1, \tag{2.53}$$

we can write

$$\int_0^a f_i(r) \mathrm{d}r = \ln \frac{J_\perp^0}{J_\parallel^0}.$$

The value of $\ln(J_\perp^0 / J_\parallel^0)$ will vary with the wavelength. The functions $f_i(r)$ are illustrated in Figure 2.12a for a diamond suspension.

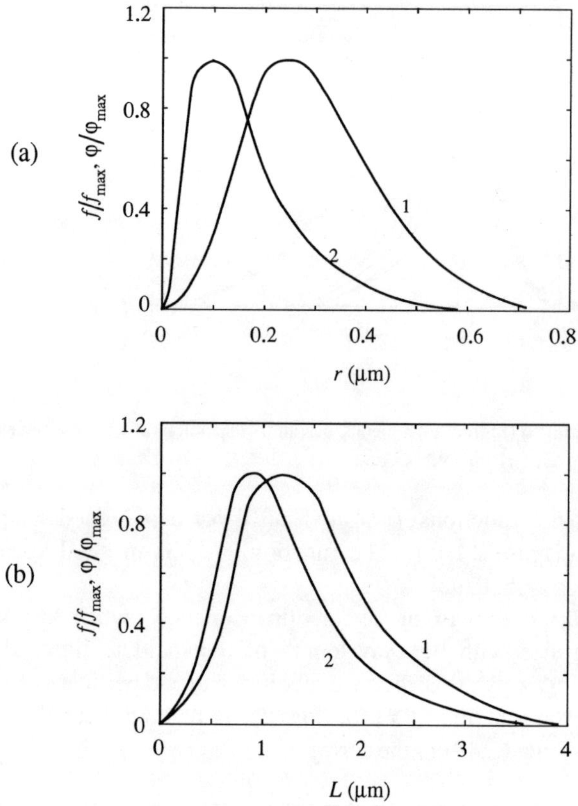


**Figure 2.10.** (a) The functions $f/f_{max}$ (curve 1) and $\varphi/\varphi_{max}$ (curve 2) for an aqueous graphite colloid. (b) The same for an aqueous palygorskit colloid.

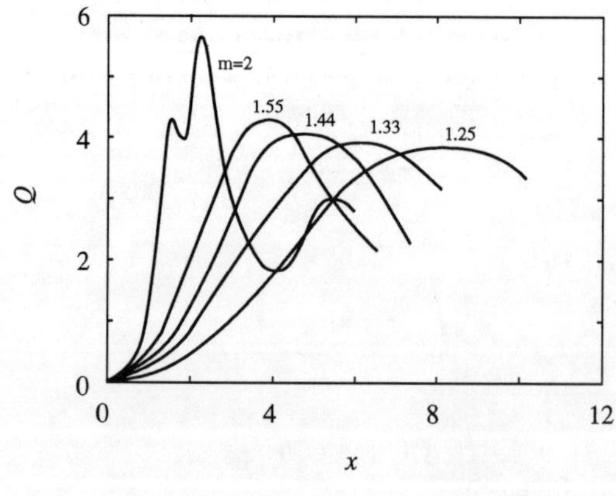


**Figure 2.11.** $Q$ versus $x$ for dielectric particles at various values of the relative refractive index $m$.

(a)

(b)

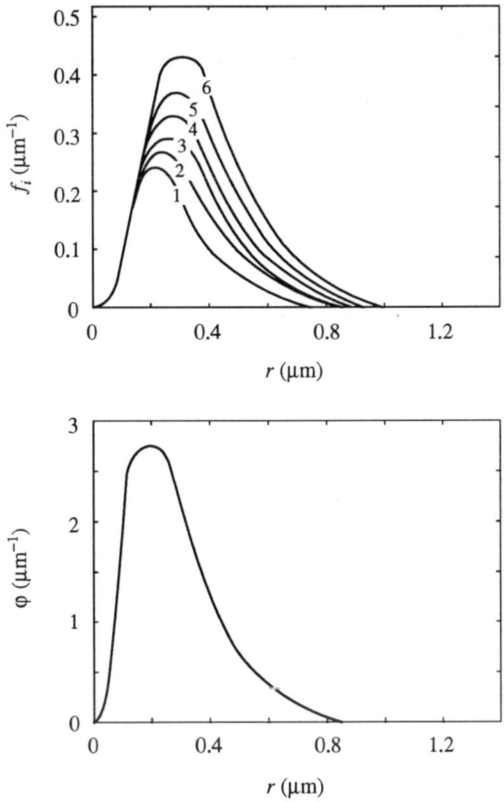


**Figure 2.12.** (a) The set of functions $f_i(r)$ for an aqueous diamond suspension. The curve numbers are equal to $i$. $\lambda_1 = 402$ nm, $\lambda_2 = 458$ nm, $\lambda_3 = 502$ nm, $\lambda_4 = 561$ nm, $\lambda_5 = 655$ nm, $\lambda_6 = 723$ nm. (b) The size distribution of particles $\varphi(r)$ calculated with the set of functions $f_i$.

With known $f_i(r)$ and equations (2.52), we can find $\varphi(r)$. Indeed, let $r/\lambda = c$, where $c$ is the chosen constant. Then, for the set $\lambda_i$ at the points $r_i = c\lambda_i$, according to (2.52), we can find the set of values:

$$\varphi(r_i) = \frac{1}{\pi c^2 F(c)} \frac{1}{\lambda_i^2} f_i(r_i). \tag{2.54}$$

By giving other values to the constant $c$, one can get other sets of $\varphi(r_i)$ with which $\varphi(r)$ can be found over the whole range of $r$. Since the factors $1/F(c_i)$ entering (2.54) are unknown, they are found from the normalization condition of (2.53) and the continuity of $\varphi(r)$.

The function $\varphi(r)$ for diamond particles obtained from the set of $f_i(r)$ in Figure 2.12a is shown in Figure 2.12b.

The above way of finding $\varphi(r)$ is suitable only when every particle in a dispersion contributes to the observable electrooptic effect. The method is not applicable to dispersions, in which the particles are either too small ($r \ll \lambda$) for their orientation to produce a change in the intensity of the linearly polarized light transmitted by the system, or too large ($r \gg \lambda$), so that their orientation equally affects the transmitted intensity at any polarization vector.

# Chapter 3

# ELECTRICAL DOUBLE LAYER OF PARTICLES IN EXTERNAL ELECTRIC FIELD

## 3.1. CLASSICAL THEORY OF ELECTROKINETIC PHENOMENA

We mentioned in Chapter 1 that anisotropy induced in a dispersion by an external electric field is associated with particle orientation, which, in conductive media, depends on the moments of force exerted on a particle by the field. The moments of force may be determined by the dipole moments of the particles and their hydrodynamic friction. So the study of particle orientation is important for the understanding of electrokinetic phenomena and electric surface effects.

Electrokinetic phenomena arise from fluid motion with respect to a particle or the vessel wall in an electric field. A systematic description of these phenomena can be found in some monographs.[45–47] Here we will outline only the aspects of the problem directly related to application of electrical orientation methods in dispersion studies.

There are two electrokinetic phenomena that influence particle behavior in external electric fields: electrical osmosis and electrophoresis. Electrical osmosis is the most common phenomenon that affects particle orientation. Electrophoresis is associated with the motion of solid particles in a suspending fluid. Emulsion droplets, gas vesicles and hard particles are involved in electrophoretic motion when an electric field is applied to the fluid. Free charges that are always present in a conductive medium are accumulated at the interfaces, causing electrical effects. The fluid resistance is not changed by electrophoresis or electrical osmosis, which indicates that the degree of molecular dissociation, and hence the fluid charge, remain the same. This fact is important for constructing particle polarization models and for the investigation of electrical conductivity of dispersions. In order to understand the behavior of a particle in an electric field, we must know the charge redistribution between the interface and the suspending fluid, or the structure of the electrical double layer.

The modern model of the double layer structure is based on the simple concepts advanced by H. Helmholtz, Y. Perrine and H. Fröundlich. In H. Helmholtz's model, an electrical double layer is a plane capacitor of one or several molecules thick (so called "Helmholtz capacitor"). One of the capacitor plates is an interface with the accumulated positive or negative charge fixed to it. An equal opposite charge is accumulated in the fluid as close to the particle surface as is allowed by the size of the fluid molecules and of the charge-carrying ions. The fluid serves as the other capacitor plate, but it is mobile and obeys all hydrodynamic laws. So if the external electric field has a tangential component, it will make one plate move relative to the other. Y. Perrine showed that no electrical osmosis arises in fluids with low dielectric permittivity because of the low dissociation rate of non-polar molecules. With growing polarity of the fluid molecules, the dielectric permittivity $\epsilon$ increases, thus increasing the degree of molecular dissociation. This, in turn, gives rise to free charges. Y. Perrine concluded that polyvalent ions with a charge opposite to that of the particle surface reduce the surface-bound charge and may even entirely change it. H. Fröundlich found that the influence of ions on the surface-bound charge grows with their adsorbance. These ideas contributed much to the understanding of the adsorption mechanism of charge accumulation at the interface. The nature of adsorption forces was clarified in the fundamental studies by B. V. Deryagin and his associates.

An alternative mechanism of the double layer formation is dissociation of ionogenic molecules. This mechanism is characteristic of biological particles and macromolecules, e.g. proteins, DNA, RNA, etc., which contain ionogenic groups, such as carboxyl aminogroups and nitric bases. Sometimes, different sites on a protein, DNA or RNA molecule have different charges, thus creating an electrical double layer of a complex configuration. In this case variation in the medium pH due to a change in the dissociation degree may considerably affect the double layer and even change the charge sign and its concentration on the particle surface or at some sites of polyions.

The electrical double layer model suggested by H. Helmholtz provided a qualitative interpretation of electrokinetic phenomena but it did not permit exact calculations of the ion concentration in the double layer.

The mathematical problem of potential and charge distributions in a double layer was first stated by L. Jouy and S. Chapman. They started with the assumption that the ion concentration in the potential field of a double layer in equilibrium can be described adequately by the Boltzmann law for the energy distribution of ions. The redistributed ions themselves create a space charge density $\rho$ which determines the potential distribution within the double layer according to Poisson's electrostatic equation. The equation for this distribution was called the Poisson–Boltzmann equation.

We present here the derivation procedure for the Poisson–Boltzmann equation described by S. S. Dukhin,[45] because it can be extended to a non-equilibrium double layer and provide a better understanding of the problem of particle polarization in an electric field.

The equilibrium condition for a double layer may be considered as an equality of two oppositely directed ion fluxes. One is defined by the potential variation $\Phi_{\text{eq}}(x)$ in the double layer with varying distance $x$ to the surface. The other flux has a diffuse character and is associated with the ion concentration gradient within the layer. The two fluxes must be compensated individually for all ion species present in the fluid. The condition for the absence of the total ion flux can be written as

$$j^i = -D_n^i \frac{\mathrm{d}C_{\text{eq}}^i}{\mathrm{d}x} + \frac{q^i}{kT} D_n^i C_{\text{eq}} \frac{\mathrm{d}\Phi_{\text{eq}}}{\mathrm{d}x} = 0, \tag{3.1}$$

where $j^i$ is the total flux of ions of the $i$-th kind, $D_n^i$ is the translational diffusion constant, $q^i$ is the ion charge, $C_{\text{eq}}^i$ is the concentration of the $i$-ions in the double layer.

The equations entering (3.1) have the following boundary conditions:

$$C_{\text{eq}}^i(x)\big|_{x\to\infty} \to C_0^i, \qquad \Phi_{\text{eq}}(x)\big|_{x\to\infty} \to 0. \tag{3.2}$$

Here $C_0^i$ is the ion concentration in the electrolyte bulk. By solving equation (3.1) with (3.2), one can get a correlation between the ion concentration and the potential distribution

$$C_{\text{eq}}^i(x) = C_0^i \exp\left(-\frac{q^i \Phi_{\text{eq}}(x)}{kT}\right). \tag{3.3}$$

It is easy now to turn from ion concentration to charge density by multiplying $C_{\text{eq}}^i$ by the ion charges and by summing up the products. So we have the following expression for the charge density distribution $\rho(x)$:

$$\rho(x) = \sum_i q^i C_{\text{eq}}^i(x). \tag{3.4}$$

In order to separate $C_{\text{eq}}^i$ and $\Phi_{\text{eq}}(x)$ entering equation (3.3), we will use the Poisson equation $\text{div}(\epsilon \boldsymbol{E}) = 4\pi\rho$, which, for a plane double layer with $\epsilon$ independent of the layer electric field, takes the form:

$$\frac{\partial^2 \Phi_{\text{eq}}}{\partial x^2} = -\frac{4\pi}{\epsilon}\rho(x). \tag{3.5}$$

Substituting (3.4) for $\rho(x)$ in (3.5), we get the Poisson–Boltzmann equation

$$\frac{\partial^2 \Phi_{\text{eq}}}{\partial x^2} = -\frac{4\pi}{\epsilon}\sum_i q^i C_0^i \exp\left(-\frac{q^i \Phi_{\text{eq}}(x)}{kT}\right). \tag{3.6}$$

The value $E = \partial\Phi_{\text{eq}}/\partial x$ may be found by integrating equation (3.6)

$$\frac{\partial \Phi_{\text{eq}}}{\partial x} = \pm\sqrt{\frac{8\pi kT}{\epsilon}\sum_i q^i C_0^i \left[\exp\left(-\frac{q^i \Phi_{\text{eq}}(x)}{kT}\right) - 1\right]}. \tag{3.7}$$

An approximate analytical solution of equation (3.7) at low values of the potential ($\Phi_{\text{eq}} < 25$ mV) may be presented as

$$\Phi_{\text{eq}}(x) = \psi_{\text{s}}\exp(-\kappa x), \tag{3.8}$$

where $\Phi_{\text{eq}} = \psi_{\text{s}}$, if $x = 0$, $\psi_{\text{s}}$ corresponds to the surface potential value:

$$\kappa = 2\left(\frac{\pi \sum_i (q^i)^2 C_0^i}{\epsilon k T}\right)^{1/2}. \tag{3.9}$$

The quantity $\kappa^{-1}$ is the Debye screening radius characterizing the distance at which the potential drops by a factor of $e$ as compared to $\psi_{\text{s}}$. For aqueous solutions of salts with monovalent ions, the value of $\kappa$ can be derived from (3.9) with the simple formula:

$$\kappa^{-1} = \frac{3.06}{\sqrt{C_0}} 10^{-8} \text{ cm}, \tag{3.10}$$

where $C_0$ is the molar ion concentration.

Expression (3.10) shows that at the phase separation boundary the double layer thickness is inversely proportional to the ion concentration in the electrolyte.

Equation (3.7) can be solved by integration in another special case unrestricted by the magnitude $\Phi_{\text{eq}}$. We mean a symmetric binary electrolyte, for which $q^i = \pm q$ $(q > 0)$ and

$$\kappa x = \ln \frac{\tanh[q\Phi_{\text{eq}}(x)/4kT]}{\tanh(q\psi_{\text{s}}/4kT)}.$$

Note that the relationship for $\kappa x$ represents a transcendental equation with respect to $\kappa$ and $\Phi_{\text{eq}}$, therefore it is reasonable to find a general solution of (3.7) with a computer using, for example, the Runge–Kutta method. For a symmetric electrolyte the relation between the surface charge density $\sigma_{\text{s}}$ and the surface potential $\psi_{\text{s}}$ may be also deduced analytically from (3.7):

$$\sigma_{\text{s}} = \frac{2\epsilon k T C_0}{\pi} \sinh\left(\frac{q\psi_{\text{s}}}{2kT}\right).$$

The Jouy–Chapman theory ignores ion sizes and their failure to be in infinite proximity to the particle surface. Stern's theory is more general and free from this limitation. It combines the Helmholtz–Smolukhovskii and Jouy–Chapman concepts. In this theory, a double layer consists of a dense sublayer and a diffuse sublayer. The dense sublayer, or Stern layer, represents a monomolecular capacitor with plates formed by ions adsorbed at the surface; the plate thickness varies with the ion size. The structure of the Stern layer is the same as that of

(a)

(b)

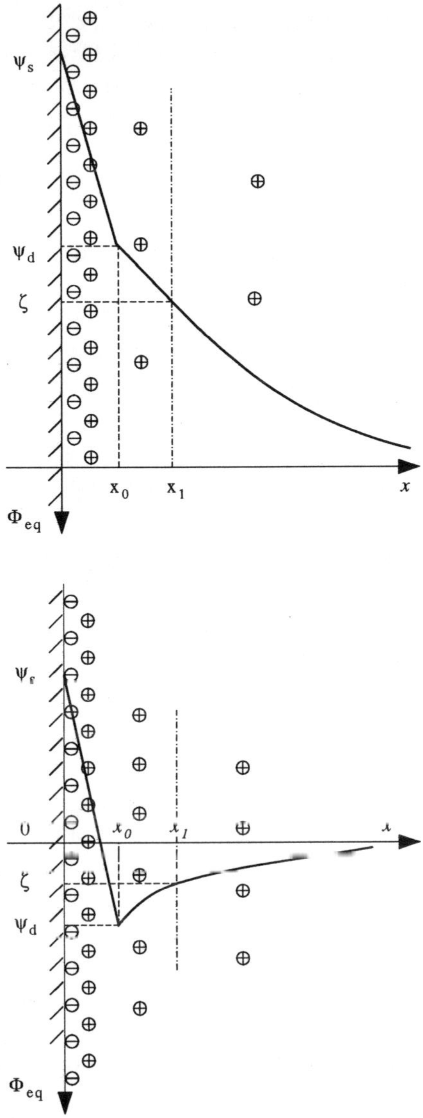


**Figure 3.1.** The double layer structure according to the Stern model.

the Helmholtz capacitor; its potential varies linearly from the value $\psi_s$ on the surface, where the ions are adsorbed, to the value $\psi_d$ at the interface between the Stern layer and the diffuse sublayer. Due to the adsorptive attraction of counterions to the surface, their concentration on the outer plate may become so high that the potentials $\psi_s$ and $\psi_d$ will have different signs, as shown in Figure 3.1a.

The diffuse double layer consists of free ions. The surface charge density $\sigma_s$ is compensated, with distance from the particle surface, by the charge density in the Stern layer $\sigma_{St}$ and in the diffuse layer $\sigma_d$. The potential $\Phi_{eq}$ of the diffuse layer is described by the Poisson–Boltzmann equation. Its variation with distance $x$ is described, at small $\psi_d$, by analogy with (3.8), i.e.

$$\Phi_{eq} = \psi_d \exp\left[-\kappa(x - x_0)\right], \tag{3.11}$$

where $x_0$ is the distance from the surface to the boundary between the Stern layer and the diffuse layer.

Expressions (3.8) and (3.11) have been derived from equation (3.5), which is valid only for a plane electrical double layer. Practically, this means that the surface curvature radius $a$ should be sufficiently large as compared with the double layer thickness: $\kappa a \ll 1$. However, the solution of the Poisson–Boltzmann equation can also be obtained for spherical and cylindrical particles. For example, for spheres of radius $a$ the solution has the form:

$$\Phi_{eq}(R) = \psi_s \frac{a}{R} \exp\left[-\kappa(R - a)\right]. \tag{3.12}$$

Here $R$ is the distance from the sphere center. Expression (3.12) has been obtained on the assumption of low potential when the inequality $q\Phi_{eq}/kT \ll 1$ is valid.

## 3.2. DOUBLE LAYER CHARACTERISTICS

The total charge of an electrical double layer of a particle is equal to zero, therefore electrophoretic motion of particles would be impossible if all double layer charges moved together with the particle. In fact, only some charges of the diffuse layer and the Stern layer charges move together with a particle. In order to separate the fluid retarded at the particle surface due to the frictional drag from the moving fluid which possesses regular bulk viscosity, the concept of slip plane has been introduced. The location of the slip plane still remains an unclear and much debatable subject. The slip plane potential is usually denoted as $\zeta$-potential. In Figure 3.1 the slip plane is represented by a dot-and-dash line parallel to the interface plane. It is often thought to coincide with the plane separating the Stern layer from the diffuse layer, i.e. $\psi_d = \zeta$. However, there are at least two reasons to believe that this assumption is wrong and that the exact calculation of the $\zeta$-potential is impossible. First, there may exist gel-like and other boundary layers on solid surfaces in polar fluids, which may possess anomalous properties due to the influence of the particle surface. Second, some surfaces are very rough, which shifts the slip plane away from the Stern layer, decreasing the value of the $\zeta$-potential.

These considerations are equally true for the slip plane position in electroosmotic motion of a fluid. The fact that there is no exact calculation of its position makes the $\zeta$-potential a semi-quantitative characteristic. However, it is useful to introduce this parameter because it is sensitive to variation in the ionic composition of electrolytes. The slip plane potential is one of the few double layer characteristics that can be measured experimentally. Its value can be found from the electrophoretic mobility of particles[48] or from the electroosmotic rate which is related to the $\zeta$-potential by Smolukhovskii's formula:

$$\zeta = \frac{4\pi\eta u}{\epsilon E}, \tag{3.13}$$

where $\eta$ is the fluid viscosity, $E$ is the electric field in the dispersion, $u$ is the particle velocity.

Expression (3.13) has been derived on the assumption of an equilibrium double layer. However, the double layer around a moving particle should not be considered as equilibrium so this expression must be supplemented by new terms. The problem of electrophoretic motion has been analyzed by S. S. Dukhin and B. V. Deryagin in detail.[48]

Another important characteristic of an electrical double layer is surface conductivity $\kappa^\sigma$ that can also be measured experimentally. Since the surface conductivity affects the current flowing in the vicinity of a particle and the particle orientation in the field, the magnitude $\kappa^\sigma$ is of importance in electrooptic and other studies involving external electric field. The value of $\kappa^\sigma$ can be found from electrical orientation studies as a fitting parameter when comparing experimental and theoretical data.

If particles are placed in an ion-containing fluid, some ions of one sign will be adsorbed on the particle surface, while some ions of the other sign will be close to the surface. The concentration of ions with both signs far from the particle surface will be lower than in the double layer. A consequence is that the tangential component of the current density along the surface will be different from that in the bulk. Therefore, the electrical conductivity of a dispersive medium differs from that of a dispersion not only due to its smaller volume and the current curvature near the particles but also to excess current density in the double layer. The excess current may be associated with ion motion in both the diffuse and Stern layer. According to Ohm's differential law, we can write:

$$I_\mathrm{s} = \kappa^\sigma E_\tau, \tag{3.14}$$

where $E_\tau$ is the external field tangential component, $I_\mathrm{s}$ is the surface current density that can be found as the difference between the bulk current densities at the particle surface, $j(x)$, and outside the double layer, $j_0$:

$$I_\mathrm{s} = \int_0^\infty [j(x) - j_0]\,\mathrm{d}x.$$

The existence of particle surface conductivity and its contribution to electrical conductivity of dispersions were pointed out by N. Smolukhovskii. In the last two decades much work has been done on correlations between $\kappa^\sigma$ and other double layer parameters.[20,45,46,49] Ion concentration considerably affects the respective contributions of the Stern and diffuse layers to the surface conductivity. For example, at a decimolar ion concentration in a dispersion, 80–90% counterions are located in the Stern double layer.[50] If, however, the ion concentration is reduced by one order of magnitude, most of the counterions will be found in the diffuse layer. One may assume that at small concentrations the surface current density is determined by the diffuse layer, then

$$I_s = \sigma_d u^i E_\tau, \tag{3.15}$$

where $u^i$ is the ion mobility in the electrolytic solution, or dispersion; $\sigma_d$ is the charge density in the diffuse layer. From comparison of (3.14) and (3.15) for low ion concentrations, we can write $\kappa^\sigma = \sigma_d u^i$.

In a simple model, where the slip plane coincides with the boundary of the Stern layer which only slightly contributes to excess current at the surface, the surface conductivity is related to the electrokinetic potential by Bickerman's formula:

$$\kappa^\sigma = \frac{2e^2}{kT}\frac{C_0}{\kappa}\left\{\left[\exp\left(-\frac{\tilde{\zeta}}{2}\right) - 1\right] D^+(1+3m^+) \right.$$
$$\left. + \left[\exp\left(\frac{\tilde{\zeta}}{2}\right) - 1\right] D^-(1+3m^-)\right\}, \tag{3.16}$$

where

$$\tilde{\zeta} = \frac{\zeta e}{kT},\ m^\pm = \left(\frac{kT}{e}\right)^2 \frac{\epsilon}{2\pi\eta D^\pm}.$$

If the double layer structure is more complex due to the surface roughness, to gel-like or anomalous layers, the slip plane is farther from the particle surface than the Stern layer boundary. In this case the surface conductivity cannot be directly related to the electrokinetic potential. We need more information about the double layer which can be obtained, for example, from the Stern layer potential. D. A. Fridrikhsberg arrived at the conclusion that the fluid corresponding to the region between the Stern layer potential $\psi_d$ and the $\zeta$-potential in Figure 3.1 is hydrodynamically motionless but conductive. That allowed him to generalize Bickerman's formula and derive an expression relating $\kappa^\sigma$, $\psi_d$ and $\zeta$:

$$\kappa^\sigma = \frac{2e^2}{kT}\frac{C_0}{\kappa}\left\{D^+\left[\exp\left(-\frac{\tilde{\zeta}}{2}\right) - 1\right](1+3m^+) + D^-\left[\exp\left(\frac{\tilde{\zeta}}{2}\right) - 1\right]\right.$$
$$\times (1+3m^-) + D_\sigma^+\left[\exp\left(-\frac{\tilde{\psi}_d}{2}\right) - \exp\left(-\frac{\tilde{\zeta}}{2}\right)\right]$$
$$\left. + D_\sigma^-\left[\exp\left(\frac{\tilde{\psi}_d}{2}\right) - \exp\left(\frac{\tilde{\zeta}}{2}\right)\right]\right\}, \tag{3.17}$$

where $D_\sigma^\pm$ are the diffusion coefficients for ions in the boundary layer, $\tilde{\psi}_\mathrm{d} = e\psi_\mathrm{d}/kT$. Thus, the quantities $\zeta$ and $\kappa^\sigma$ are independent and important kinetic characteristics of a double layer, and they should be measured simultaneously to make its description complete. With these data and using (3.16) and (3.17), one can find the difference between $\psi_\mathrm{d}$ and $\zeta$ and determine the distance between the Stern layer boundary and the slip plane.

## 3.3. INDUCED DIPOLE MOMENT OF PARTICLES

Investigation of electrooptic effects and conductivity can provide an important characteristic of dispersed material, namely particle polarizability, which is related to the electrical characteristics of the particle surface and the dispersion as a whole.

The initial model of particle polarizability dealt only with polarization and electric current flowing through a particle and the medium, taking into account their electrical conductivities and permittivities. In recent years, theory of particle polarizability has been considerably extended to include interface effects and allows us to relate the induced dipole moments of particles and the characteristics of the electrical double layer. It has been shown that the double layer polarization makes a major contribution to the polarizability of colloidal particles in polar dispersive media.[47,51] The development of electrooptic and conductometric techniques opens up an opportunity for experimental testing of particle polarizability models and for determination of double layer characteristics.

Real dispersions are in most cases polydisperse and polymorphous, so a polarizability theory can be verified reliably only by methods that take into account polydispersity and polymorphism. It will be shown in Chapter 4 that polydispersity plays a decisive role in steady-state electrooptic effects, while the influence of particle shapes has been demonstrated by orientation studies to be slight. The point is that processes inducing electrical orientation are largely determined by the electric current and hydrodynamic flows around a particle, but the flow spatial parameters are not very sensitive to small variations in the particle shape. This is not the case, however, for purely optical properties of dispersions, when even a slight variation in the particle shape changes drastically the light scattering by a particle. If conductometric or electrooptic techniques do not use the light scattering diagrams, a statistical ensemble of irregular particles can be replaced by an ensemble of ellipsoids of revolution. The obtained double layer characteristics are then compared to each other and to the data obtained by other methods, for instance, by electrokinetic measurements.

The polarizability theory for colloidal particles was further developed by S. S. Dukhin and V. N. Shilov in the 1970–1980s. They did a detailed study of the influence of the double layer characteristics on the induced dipole moment.

Let us now consider the basic theoretical results necessary for the calculation of colloidal particle orientation in an electric field. If the dispersive fluid

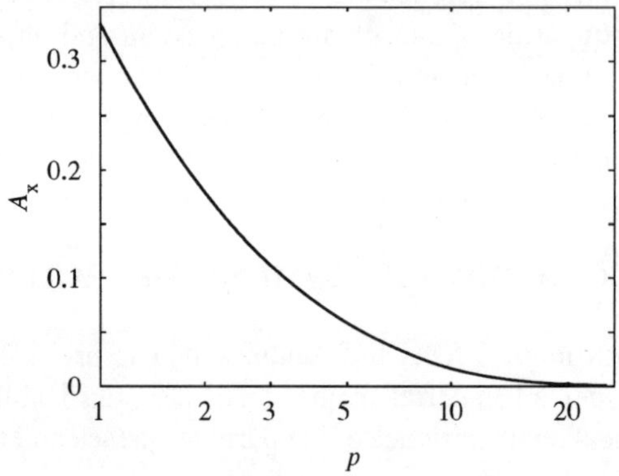


**Figure 3.2.** Depolarization factor $A_x$ as a function of the spheroid semiaxis ratio.

and the particle material do not possess electrical conductivity, the polarizability of particles of a given size is determined only by the permittivities of the fluid and the material. For a sphere the polarizability $\gamma_{\text{sph}}$ can be expressed as

$$\gamma_{\text{sph}} = r^3 \epsilon \, \frac{\epsilon' - \epsilon}{2\epsilon + \epsilon'},$$

where $\epsilon$ is the permittivity of the fluid, $\epsilon'$ is the permittivity of the particle material, $r$ is its radius. For ellipsoidal particles, the polarizability is also determined by their orientation with respect to the electric field vector. If the ellipsoid axis $x$ is oriented along the field, the polarizability is described by

$$\gamma_x = \frac{\epsilon V}{4\pi} \frac{\epsilon' - \epsilon}{\epsilon + (\epsilon' - \epsilon) A_x}, \tag{3.18}$$

where $V$ is the ellipsoid volume, $A_x$ is the depolarization factor along the $x$-axis. The depolarization factors for the other two axes, $y$ and $z$, are the functions of the semiaxis ratio only and are related as

$$A_x + A_y + A_z = 1. \tag{3.19}$$

For ellipsoids of revolution, along the $x$-axis we have:

$$A_y = A_z = \frac{1 - A_x}{2}. \tag{3.20}$$

The dependence of $A_x$ on the semiaxis ratio $p$ for a spheroid has the form:

$$A_x = \int_0^\infty \frac{p^2 b^3}{2(p^2 b^2 - S)^{3/2}(b^2 + S)} \mathrm{d}S. \tag{3.21}$$

It is presented in Figure 3.2.

It follows from (3.18) and (3.20) that the polarizability of a spheroid along its short axis is defined as:

$$\gamma_y = \gamma_z = \frac{\epsilon V}{4\pi} \frac{\epsilon' - \epsilon}{\epsilon + (\epsilon' - \epsilon)(1 - A_x)/2}. \tag{3.22}$$

Formulas (3.18) and (3.22) show that the value $\gamma_y = \gamma_z$ for a spheroid is smaller than $\gamma_x$, therefore the polarizability anisotropy $\gamma = \gamma_x - \gamma_y$ always exceeds zero, i.e. the particles are oriented with their long axes along the field.

The polarizability of a conductive particle in a non-conductive medium may be described in much the same way as for a non-conductive particle with an infinitely large permittivity. Taking $\epsilon/\epsilon' = 0$ in (3.18), we obtain the polarizability expressions for a conductive particle in a non-conductive medium:

$$\gamma_x = \frac{\epsilon V}{4\pi A_x}, \qquad \gamma_y = \frac{\epsilon V}{2\pi(1 - A_x)}, \qquad \gamma = \frac{\epsilon V}{4\pi} \frac{1 - 3A_x}{A_x(1 - A_x)}.$$

Similarly, assuming $\epsilon'/\epsilon = 0$ in (3.18) and (3.22), we obtain expressions for the polarizabilities $\gamma_x$ and $\gamma_y$, and the anisotropy of polarizability $\gamma$ for non-conductive particles in a conductive medium:

$$\gamma_x = \frac{\epsilon V}{4\pi(A_x - 1)}, \qquad \gamma_y = \frac{-\epsilon V}{2\pi(1 + A_x)}, \qquad \gamma = \frac{\epsilon V}{4\pi} \frac{1 - 3A_x}{1 - A_x^2}.$$

If both the fluid and the dispersed material are conductive, the polarizability is defined by the relationship:

$$\gamma_x = \frac{\epsilon V}{4\pi} \frac{K'/K - 1}{1 + (K'/K - 1)A_x}, \tag{3.23}$$

where $K'$ is the conductivity of the material; $K$ is the conductivity of the fluid. Relationship (3.23) is valid when the double layer concentrational polarization, to be considered in Sections 3.4 and 3.5, is inessential.

One can see from (3.23) that at $K' > K$ the polarizability $\gamma_x$ is positive and the induced dipole moment is along the field; at $K' < K$ it is reversed. Of special interest is the case of a non-conductive particle with a surface conductivity due to the surrounding double layer. Here we cannot use (3.23) because of the difference in the conductivities of the particle, fluid and double layer. Since the double layer conductivity varies with the distance to the particle surface and with the current vector, we need such a theory that would allow for the double layer polarization under the external field. Below, we will analyze this situation.

## 3.4. ISOPOLARITY OF A PARTICLE

When studying the influence of an electric double layer on an induced dipole moment, we should primarily discuss the contribution of the surface conductivity $\kappa^\sigma$, which is due to ion motion in the layer under the external field and, therefore, responsible for the double layer polarizability.

The potential distribution around a particle is represented as the sum:

$$\Phi(R,\theta) = \Phi_0(R,\theta) + \Phi_p(R,\theta),$$

where $R$ and $\theta$ are spherical coordinates with $\theta$ measured from the electric field vector. The first summand is responsible for $\Phi(R,\theta)$ if $\kappa^\sigma = 0$. It can be obtained by solving the equation:

$$\nabla^2 \Phi_0 = 0 \tag{3.24}$$

under the condition that on the particle surface

$$\frac{\partial \Phi_0}{\partial R} = 0. \tag{3.25}$$

For a sphere of radius $a$ at $\kappa a \gg 1$ and $R \gg a$, the asymptotes of the solution can be represented as

$$\Phi_0(R,\theta) = -E\left(R + \frac{a^3}{2R^2}\right)\cos\theta,$$

where the zero potential is the potential at the particle center.

The following reasoning can be employed for finding the potential $\Phi_p$. The electric current distribution over a particle surface is non-uniform: the current is zero at the particle's poles, where the external field is normal to the surface, and is maximum at the sides, where the external field is parallel to it. For a sphere, the surface current maximum lies at the equator ($\theta = 90°$). Current variation along the surface, without charge accumulation, in the steady-state case is provided by ion motion in the direction normal to the surface. This condition cannot be expressed in a way similar to (3.25) and be valid for $\Phi_0$. The formulation of the problem of the potential $\Phi_p$ involves the Laplace equation and the following boundary condition:

$$K \left.\frac{\partial \Phi}{\partial R}\right|_s = \mathrm{div}_s\left(\kappa^\sigma \frac{\partial \Phi}{\partial \theta}\right), \tag{3.26}$$

where $\mathrm{div}_s$ is the surface divergence, $\left.\frac{\partial \Phi}{\partial R}\right|_s$ is the normal component of the electric field at the particle surface.

This approach permits the calculation of the potential and dipole moment of a simple particle.[20] For a sphere, the dipole moment can be found from

$$d = -\frac{1}{2}a^3 E\left(1 - 3\frac{\mathrm{Rel}}{1 + \mathrm{Rel}}\right)\epsilon, \tag{3.27}$$

where $\mathrm{Rel} = \kappa^\sigma / Ka$ is a dimensionless parameter used to describe the contribution of surface conductivity to the polarizability of a non-conductive particle in a conductive medium.

Relation (3.27) shows that the dipole moment may have different signs, depending on the parameter Rel. For instance, at small Rel, when the surface conductivity practically vanishes, the dipole moment is antiparallel to the field. At large $\kappa^\sigma$, it is directed along the field, which is due to the excess charge accumulated at the particle poles, when the excess current flows along the surface. Thus, a particle with a large charge is polarized along the field. At some values of surface conductivity the dipole moment is equal to zero, and this state is known as isopolarity.

It follows from (3.27) that a spherical particle is not polarized at $\mathrm{Rel}_{\mathrm{iso}} = 0.5$. As shown by S. S. Dukhin[45] and D. A. Fridrikhsberg,[49] $\kappa^\sigma$ reveals only a several-fold increase, when the electrical conductivity $K$ changes by a factor of several tens or even hundreds. This means that the dipole moment changes largely due to the changing conductivity of the dispersive medium. Note that the parameter Rel should be introduced only for a plane double layer ($\kappa a \gg 1$), i.e. only for strongly charged particles, when the ion concentration in the fluid is high and the Debye screening radius is small. If the Stern and electrokinetic potentials are large and the ion mobility at the surface is the same as in the fluid, the value of Rel is described by a simple expression:[20]

$$\mathrm{Rel} = \frac{1}{\kappa a}\left[\exp\left(\frac{\tilde{\psi}_{\mathrm{d}}}{2}\right) + 3m\exp\left(\frac{\tilde{\zeta}}{2}\right)\right]. \tag{3.28}$$

However, the introduction of the parameter Rel and employment of formula (3.28) are useful only for monodisperse systems, since the parameter varies greatly with the particle size. For a polydisperse system it is reasonable to introduce a dimensional parameter $Z = \kappa^\sigma / K$. An expression similar to (3.28) will be:

$$Z = \frac{1}{\kappa}\left[\exp\left(\frac{\tilde{\psi}_{\mathrm{d}}}{2}\right) + 3m\exp\left(\frac{\tilde{\zeta}}{2}\right)\right]. \tag{3.29}$$

The analysis of the surface conductivity contribution to particle polarizability has been based on the assumption (3.26) that the total charge is not accumulated on the surface. However, this condition is insufficient for a real system containing multivalent ions of different signs. In a steady-state case, variation in the surface current must be compensated by fluxes of all sorts of ions in the direction normal to the particle surface. If the relative concentration of ions of different sorts varies with distance to the particle surface, there will be a redistribution of charge density near the particle, producing additional (concentrational) polarization of the double layer. This polarization mechanism can be understood from analysis of space distribution of ionic fluxes in a steady-state case.[20] One should note that concentrational polarization does not affect relations (3.28) and (3.29) but will change the expression for the dipole moment. For instance, for a sphere, instead of (3.27), we have:

$$d = -a^3 E \left( \frac{1}{2} - \frac{3\mathrm{Rel}}{1 + 4\mathrm{Rel}} \right) \epsilon, \tag{3.30}$$

where the quantity Rel is defined by (3.28).

Lately, successful efforts have been made to develop a theory of double layer concentrational polarization in a strong electric field with account of electrophoretic motion of particles.[52–55]

## 3.5. FREQUENCY DEPENDENCE OF INDUCED DIPOLE MOMENT

We have so far discussed polarization of particles in a static electric field under steady-state conditions. Of greater interest, however, are polarization studies in sine-shaped fields of different frequencies. They allow one to avoid some side effects, such as electrophoretic orientation of particles and field distortions due to the electrode polarization during the measurements. In order to find out at what frequencies relations (3.18), (3.19) or (3.30) are valid, one must know the relaxation time of a free polarization charge of a particle. It is described as[20]

$$\tau_\kappa = \frac{1}{\kappa^2 D_\mathrm{e}}, \tag{3.31}$$

where

$$D_\mathrm{e} = \frac{D^+ D^- (z^+ + z^-)}{z^+ D^+ + z^- D^-}.$$

Therefore, the value of circular field frequency, at which the conductivities of the medium and the particles become important, satisfies the inequality:

$$\omega \ll \kappa^2 D_\mathrm{e}. \tag{3.32}$$

At frequencies satisfying this inequality, one can describe a double layer, using the Boltzmann law for the charge distribution in a diffuse layer.

At frequencies of the order of $1/\tau_\kappa$, there is interface polarization or Maxwell–Wagner polarization. The frequency range is defined by the time of formation of a free polarization charge. The Maxwell–Wagner polarization can be described using a complex permittivity $\epsilon^*$, which allows one to combine the equations:

$$\text{div}(\epsilon \text{grad}\Phi) = -4\pi\rho \tag{3.33}$$

$$\text{div}(K \text{grad}\Phi) = \frac{\partial \rho}{\partial t}. \tag{3.34}$$

For a sine-shaped field the time variations of the potential $\Phi$ and the charge density $\rho$ with frequency $\omega$ are described by a sine-function. Then the set of equations (3.33) and (3.34) can be reduced to one equation:

$$\text{div}\left[\left(\epsilon - \frac{4\pi \text{i} K}{\omega}\right) \text{grad}\Phi\right] = 0. \tag{3.35}$$

This equation coincides with the Laplacian equation if the permittivity of a conductive medium is assumed to be a complex quantity according to the expression:

$$\epsilon^* = \epsilon - \frac{4\pi \text{i} K}{\omega} \tag{3.36}$$

If there is no excess charge at the interface, then:

$$\epsilon^* \left.\frac{\partial \Phi}{\partial n}\right|_s = \epsilon'^* \left.\frac{\partial \Phi'}{\partial n}\right|_s .$$

and the dipole moment of a particle is described in the same way as for a particle with the permittivity $\epsilon'$ in a medium with the permittivity $\epsilon$. The only difference here is that these permittivities represent complex quantities. By analogy with (3.18) and (3.22), we can write for the polarizability of an ellipsoid of revolution.

$$\gamma_x^* = \frac{\epsilon V}{4\pi} \frac{\epsilon'^* - \epsilon^*}{\epsilon^* + (\epsilon'^* - \epsilon^*)A_x} \tag{3.37}$$

$$\gamma_y^* = \gamma_z^* = \frac{\epsilon V}{4\pi} \frac{\epsilon'^* - \epsilon^*}{\epsilon^* + (\epsilon'^* - \epsilon^*)(1 - A_x)/2}. \tag{3.38}$$

At low frequencies the polarizability of a particle can be described by expressions (3.37) and (3.38) with the substitutions $\epsilon'^* \to K'$ and $\epsilon^* \to K$. In this case the frequency $\omega$ must satisfy the inequality:

$$\omega \ll \frac{K' + A_x(K' - K)}{\epsilon + A_x(\epsilon' - \epsilon)}.$$

At high frequencies, when

$$\omega \gg \frac{K' + A_x(K' - K)}{\epsilon + A_x(\epsilon' - \epsilon)},$$

the substitutions $\epsilon'^* \to \epsilon'$ and $\epsilon^* \to \epsilon$ should be made in the same expressions.

Formulas (3.37) and (3.38) for the Maxwell–Wagner mechanism of particle polarizability ignore the contribution of surface conductivity. In a simple case $\kappa^\sigma$ is taken into account when it is independent of frequency; for this the substitutions[27,56] $\epsilon'^* \to K' + 2\kappa^\sigma/a$ and $\epsilon^* \to K$ in (3.37) and (3.38) are quite sufficient.

When $\kappa^\sigma$ varies with the field frequency, one should take into account the mechanism of double layer concentrational polarization to describe the frequency dependence of polarizability.[57] The influence of concentrational polarization on the induced dipole moment of a sphere in weak fields is such that the value of $d$ can be calculated by formula (3.27) and at low frequencies by (3.30).

The relaxation time of concentrational polarization of a double layer is defined as $\tau_c = a/2D_e$, therefore, the concentrational polarization will change at frequencies close to

$$\omega \approx \frac{D_e}{a^2}. \tag{3.39}$$

Expressions (3.31) and (3.39) are close enough, if the double layer thickness is comparable with the particle size ($\kappa a \approx 1$). At $\kappa a \gg 1$ (double layer is plane), there are two regions of polarizability change. If the inequality $\kappa a \gg 1$ is not fulfilled, these regions overlap and their separation becomes a very difficult problem.

## 3.6. POLARIZABILITY OF PARTICLES WITH AXIAL SYMMETRY

The polarizability of a cylinder of length $l_c$ and diameter $2a$ oriented perpendicular to the field can be described by:[20]

$$\gamma_\perp = \frac{a^2 l_c}{2}\left(1 - \frac{\mathrm{Rel}_\perp}{1/2 + \mathrm{Rel}_\perp}\right)\epsilon, \tag{3.40}$$

where $\mathrm{Rel}_\perp = 2\kappa^\sigma/Kd$ is determined by the particle diameter and is independent of its length. Expression (3.40) resembles that for a sphere. The difference, however, is that at low frequencies and any $\mathrm{Rel}_\perp$ the dipole moment is antiparallel to the field, in contrast to a sphere, and isopolarity becomes unattainable.

When a particle is oriented along the field,[20] the surface current is constant over the entire side surface, so we can ignore the ion flux component normal to the surface. The lateral effects on $\kappa^\sigma$ can be neglected if the particle length greatly exceeds its diameter. In this case the charges of one sign are accumulated at one end of the particle, creating a total charge $Q^-$, while the charge $Q^+ = Q^-$ is accumulated at the other end. The values of these

**Table 3.1**

| | | | | | | |
|---|---|---|---|---|---|---|
| $\text{Rel}_{\|\|}$ | $5 \times 10^{-6}$ | $2.5 \times 10^{-5}$ | $5 \times 10^{-5}$ | $10^{-4}$ | $5 \times 10^{-4}$ | 0.025 |
| $\mu$ | 1 | 0.99 | 0.94 | 0.9 | 0.89 | 0.81 |
| $\text{Rel}_{\|\|}$ | 0.01 | 0.25 | 0.1 | 0.2 | 0.3 | 0.4 |
| $\mu$ | 0.64 | 0.42 | 0.18 | 0.09 | 0.08 | 0.067 |

charges vary with $\kappa^{\sigma}$, and the polarizability of a cylinder can be approximately described as follows:

$$\gamma_{||} = -\frac{a^2 l_c}{4}\left(\frac{1}{3} - \frac{\pi}{4}\text{Rel}_{\perp}\right)\epsilon. \tag{3.41}$$

Expression (3.41) is rigorous only for a very elongated spheroid. The difference between the polarizabilities of a very elongated spheroid and a cylinder is small and determined by the fact that in a spheroid the contribution of ion fluxes normal to the surface to $Q^+$ and $Q^-$ increases towards the particle ends. A more rigorous expression for a cylinder is[20]

$$\gamma_{||} = \frac{a^2 l_c}{4}\left(\frac{1}{3} - \frac{\pi}{4}\mu\left(\text{Rel}_{||}\right)\text{Rel}_{\perp}\right)\epsilon, \tag{3.42}$$

where $\text{Rel}_{||} = 4a^2/l_c^2$ and the correction factor $\mu$ depends on Rel (see Table 3.1). Expression (3.42) gives a somewhat larger value of $\gamma_{||}$ than equation (3.41).

Let us consider the polarizability of a spheroidal particle. At high frequencies, when the inequalities

$$\omega \gg \frac{2D_e}{a^2}, \qquad \omega \gg \frac{2D_e}{b^2}$$

are valid and $a$ and $b$ are the long and short semiaxis, respectively, the particle polarization can be described as a Maxwell–Wagner one with expressions (3.37) and (3.38), ignoring the concentrational polarization. We cannot at present take into account the concentrational polarization for spheroids in a sine-shaped field of arbitrary frequency because of great mathematical difficulties. The polarizability at low frequency, when the concentrational polarization takes a steady-state value,[20] will be considered below. Expressions for $\gamma_{||}$ and $\gamma_{\perp}$ have the form:

$$\gamma_{||} = \frac{\epsilon r^3}{3}\left(\frac{3}{2}I_{||}\text{Rel} - 1\right)\left[1 + A_x\left(\frac{3}{2}I_{||}\text{Rel} - 1\right)\right]^{-1} \tag{3.43}$$

$$\gamma_{\perp} = \frac{\epsilon r^3}{3}\left(\frac{3}{2}I_{\perp}\text{Rel} - 1\right)\left[1 + \frac{1 - A_x}{2}\left(\frac{3}{2}I_{\perp}\text{Rel} - 1\right)\right]^{-1}, \tag{3.44}$$

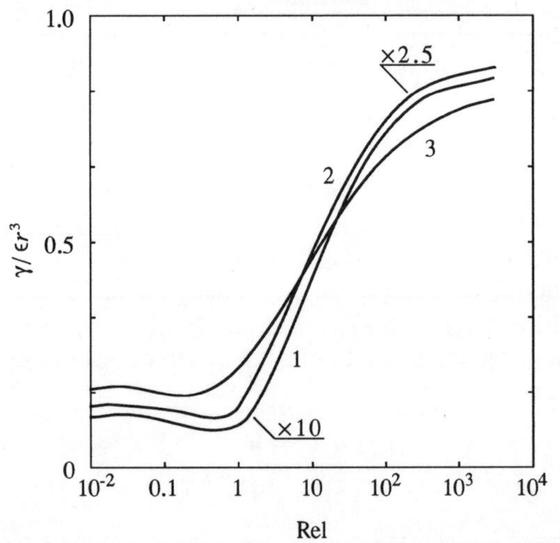


**Figure 3.3.** The plots of $\gamma/\epsilon r^3$ versus Rel for spheroids at $p = 1.5$ (curve 1), $p = 3.0$ (curve 2), $p = 5.0$ (curve 3).

where $r$ is the radius of a sphere of the same volume as the particle.

It is now more convenient to use $\mathrm{Rel} = \mathrm{Z/b} = \mathrm{Zp}^{1/3}/\mathrm{r}$ ($p$ is a ratio of the spheroid semiaxes, $p = a/b$). The coefficients $I_{\|}$ and $I_{\perp}$ are defined by the spheroid elongation and related to $p$ by the following expressions

$$I_{\|} = c\left[\arcsin\left(\frac{1}{c}\right)(2 - c^2) + \sqrt{c^2 - 1}\right], \quad c = \frac{p}{\sqrt{p^2 - 1}}. \tag{3.45}$$

$$\begin{aligned} I_{\perp} &= \frac{B}{c\sqrt{c^2 - 1}} - c(B + 1)\left[\frac{1}{c^2 - 1} - \arcsin\left(\frac{1}{c}\right)\right] \\ &+ c\left[\frac{1}{2}\sqrt{c^2 - 1} + \frac{c^2}{\sqrt{c^2 - 1}} - \frac{3}{2}c^2 \arcsin\left(\frac{1}{c}\right)\right] \\ B &= \frac{p^2 + 1}{p^2 - 1}. \end{aligned} \tag{3.46}$$

The anisotropy of electrical polarizability $\gamma = \gamma_{\|} - \gamma_{\perp}$, determining the particle orientation in the electric field, first decreases with increasing Rel (Figure 3.3) from the value

$$\gamma\Big|_{\mathrm{Rel}=0} = \frac{\epsilon r^3}{3}\,\frac{1 - 3A_{\|}}{1 - A_{\|}^2} \tag{3.47}$$

(dielectric particle in a conductive medium), then passes through a minimum and then increases up to

$$\gamma\Big|_{\mathrm{Rel}\to\infty} = \frac{\epsilon r^3}{3}\,\frac{1 - 3A_{\|}}{A_{\|}(1 - A_{\|})} \tag{3.48}$$

(conductive particle in a non-conductive medium).

The calculations have shown the position of $\gamma$ minimum to vary weakly with the spheroid shape: when the semiaxis ratio changes from 1 to 5, the value of Rel corresponding to $\gamma$ minimum, changes from 0.29 to 0.32, or within 10%. This fact is very important for determination of $\gamma(\text{Rel})$ minimum for particles of varying shape. The position of the minimum depends on the spheroid semiaxis ratio and, together with (3.46)–(3.48), determines the $\gamma$ range. Within this range, $\gamma$ remains positive, and at any value of $Z$ the particles are oriented with their long axes along the field. As a consequence, such effects as electrical dichroism or conductivity anisotropy do not change their sign with changing experimental conditions. The value of $\gamma$ derived from (3.43) and (3.44) allows one to find the moment of forces acting on a particle with a double layer

$$M_f = (\gamma_{\parallel} - \gamma_{\perp})\frac{E^2 \sin 2\theta}{2},$$

which is necessary for finding the angular distribution function $\mathcal{W}(\theta)$ of the particle axes, according to (1.11). But in a general case one should take into account the hydrodynamic moment of forces associated with electroosmotic sliding of the fluid. The expressions for this contribution to the total moment of forces are now available only for very elongated and strongly charged particles.[20]

# Chapter 4

# ANISOTROPY OF ELECTRICAL POLARIZABILITY OF DISPERSED PARTICLES

## 4.1. INTRODUCTION

Electrooptic methods based on the study of field-induced light scattering anisotropy are frequently used to get information on the properties of the disperse phase. Much contribution to the development of such techniques has been made by B. R. Jennings and his British co-workers and by S. P. Stoylov together with his Bulgarian colleagues. These techniques are applicable to dispersions with particles much smaller than the incident light wavelength or shaped as thin rods or disks. A common feature of the methods is that the underlying theory describes only a monodisperse colloidal system, which restricts their applicability. Therefore, the data they provide on some real colloidal systems should be regarded as no more than approximate.

The electrooptic method described in Chapter 2 for the study of anisotropy of electrical polarizability is based on field-induced dichroism of a polydisperse colloidal system and does not have these limitations. It can be used to study real dispersions, in which particle sizes are comparable to the incident polarized light wavelength.

In this Chapter we discuss electrooptic phenomena associated with dichroism induced in a dispersion by an applied electric field. We draw a comparison between experimental data and the theoretical results by S. S. Dukhin and V. N. Shilov concerning electrical polarizability of spheroidal particles. Electrical double layer parameters are found and compared with data obtained by other techniques.

## 4.2. EXPERIMENTAL DETERMINATION OF POLARIZABILITY ANISOTROPY IN POLYDISPERSE COLLOIDAL SOLUTIONS

Electrooptic studies of dispersions involving correlations between electrical dichroism and the applied sine-shaped electric field (see (2.8)) may provide values of polarizability anisotropy for single particles and their experimental variations with the particle size and physical-chemical properties of the dispersive medium. As was pointed out in Chapter 1, it is profitable to carry out polarizability studies in alternating fields of several kilocycles per second and higher. Particle orientation in high-frequency fields is totally due to the polarizability anisotropy, because the permanent dipole moment exerts no effect whatsoever.

It is of interest to compare experimental data for irregular elongated particles and available polarizability theory for spheroids (see Chapter 3). This comparison may help us find the range of applicability of this theory to electrooptic experiments with real dispersions containing particles of varying shape and size. We may also better understand the real value of electrooptic methods for studies of electrical characteristics of dispersions.

In the polarizability theory for spheroids, the dependence of an electrooptic effect on the electric field can be calculated from the known correlation between the polarizability anisotropy and the sizes of particles of a given shape, when the polydispersity of a system is given by the dichroic-weighted size distribution function $f(r)$. Here $r$ is the radius of a sphere equivalent in volume to a spheroid which approximates a real particle. The spheroid, in turn, is considered to be equivalent to a real particle, if both have the same rotational diffusion coefficient $D$. However, $D$ does not prescribe yet the spheroid geometry, since the same value of $D$ may correspond to a great variety of combinations of volume and shape (semiaxis ratio $p$). For this reason, the shape of a spheroid should be found from experimentally observed shapes of real particles, for example, with an electron microscope. Evidently, this way of determining $p$ is very inaccurate. It will be shown later, however, that electrical characteristics of particles found from electrooptic experiments vary but slightly with $p$. This means that a monomorphic approximation is quite justifiable in electrooptic studies.

Formulas (3.43) and (3.44) for polarizability of spheroids can be used, together with the function $f(r)$, to calculate $N(E^2)$. Comparison of these calculations with the respective experimental data can clarify the question whether the polarizability theory for spheroids can provide a valid description of electrooptic effects in real dispersions. The following points should be tested experimentally:

(i) Can theoretical and experimental curves for $N(E^2)$ coincide? The shape of a theoretical curve depends on the form of the function $\gamma = \gamma(r)$ given by the polarizability theory for spheroids and is defined by the value of $Z$. (Polarizability theory for colloidal particles generally uses Rel. In polydisperse systems, however, the polarizability of particles of various sizes should be calculated for varying Rel, so it is reasonable to use the parameter $Z = \kappa^\sigma / K$, the same for all particles of a system.)

(ii) Does the minimum of theoretical $\gamma(Z)$ coincide with that in experiment? The point is that decreasing $Z$ in theory corresponds to increasing electrolyte concentration in a real dispersion.

These questions can be answered by comparing theoretical and experimental values for $\left.\frac{dN}{dE^2}\right|_{E=0}$.

In order to test experimentally the polarizability theory for spheroids, we have used aqueous colloidal dispersions of synthetic diamond, anisaldazine, anisilidenbenzidine, benzopurpurine and palygorskit.

**Preparation of Dispersions.** Diamond and palygorskit suspensions were obtained by ultrasound mixing of fine synthetic diamond powder or palygorskit paste in bidistilled water with an electrical conductivity less than a few units of $M\Omega^{-1}\ cm^{-1}$. Anisaldazine and anisilidenbenzidine colloidal solutions were prepared using a solvent substitution procedure. The material was dissolved in a small amount of ethyl alcohol, and a few droplets of the solution were added to a large volume of water at room temperature. When the alcohol solution is mixed with water, the material is crystallized to produce a colloidal solution. If the alcohol solution is cooled prior to mixing, the size of anisaldazine or anisilidenbenzidine particles is somewhat smaller. The obtained colloidal dispersions were then centrifuged to remove large inclusions and aggregates. The benzopurpurine colloidal system was prepared by directly dissolving the material in water. The number densities of particles in all the dispersions were less than $10^8\ cm^{-3}$.

If the quantity $N$ is defined only by the anisotropy of polarizability, formula (2.26) can be re-written as

$$N = \int_0^\infty f(r)\mathcal{F}(\gamma, E)\mathrm{d}r, \tag{4.1}$$

where

$$\mathcal{F}(\gamma, E) = \frac{3}{2}\frac{\int_0^1 x^2 \exp(\gamma E^2 x^2/2kT)\mathrm{d}x}{\int_0^1 \exp(\gamma E^2 x^2/2kT)\mathrm{d}x} - \frac{1}{2}. \tag{4.2}$$

Relation (4.1) is obtained by substituting $\lambda = 0$ into (2.10), $f(r)$ is the dichroic-weighted size distribution function for a given colloidal solution. The procedure for determining this function was described in Chapter 2.

Figure 4.1 presents the experimental and theoretical plots for $N(E^2)$ in aqueous colloidal dispersions of synthetic diamond, anisaldazine and anisilidenbenzidine, together with the respective size distribution functions $f(r)$ (dashed curves). The values of $Z = \kappa^\sigma/K$, at which the experimental and theoretical dependences on the particle size $r$ coincide, were computed using the least square procedure and found to be $1.6 \times 10^{-4}$, $1.3 \times 10^{-4}$, and $8.2 \times 10^{-5}$ cm, respectively. The shapes of the $N(E^2)$ theoretical curves strongly depend on $Z$, which indicates the validity of the above method based on electrooptic studies. The best fit between the calculated and experimental $N(E^2)$ is at $p = 1.2$ for anisilidenbenzidine and at $p = 1.25$ for anisaldazine and diamond. These values of $p$ are consistent with electron microscopic observations of the colloidal particles. The maximum deviations of the calculated values of $N$ for the parameters $Z$ and $p$ from the experimental ones were 2, 5, and 6% for diamond, anisaldazine, and anisilidenbenzidine, respectively; they did not exceed the measurement error for $N$. This testifies to applicability of the spheroid polarization model to both qualitative and quantitative electrooptic studies of an ensemble of numerous irregular particles. The electrical conductivity $K$ of the dispersive fluids were a few units of $M\Omega^{-1}\ cm^{-1}$.

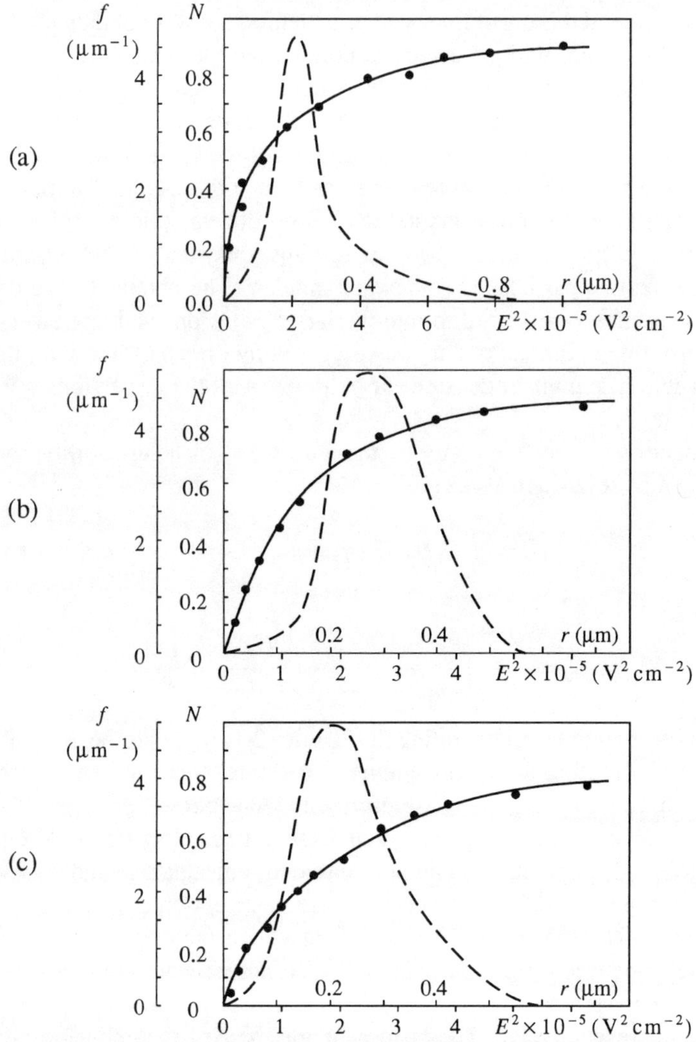


**Figure 4.1.** (a) The plots of $N(E^2)$ (●, experiment; solid curve, theory) and $f(r)$ (dashed curve) for aqueous colloidal systems of synthetic diamond. (b) The same for anisaldazine. (c) The same for anisilidenbenzidine.

The values of $\kappa^\sigma$, which are several tens of $G\Omega^{-1}$ for the particles under study, can be estimated from the derived $Z$ and $K$. These values are somewhat lower than those obtained from a surface conductivity study of palygorskit aqueous colloidal dispersions.[58] A detailed analysis of surface conductivity of colloidal particles will be given in Section 4.2. The obtained values of $Z$ permit the calculation of polarizability anisotropy for spheroids of specific sizes. Table 4.1 lists the results of the calculations.

**Table 4.1**

| Dispersion | $Z \times 10^4$, cm | $p$ | $r$, $\mu$m | $\gamma$, cm$^3$ |
|---|---|---|---|---|
| Diamond in water | 1.6 | 1.25 | 0.50 | $1.6 \times 10^{-12}$ |
| | | | 0.25 | $2.6 \times 10^{-13}$ |
| | | | 0.05 | $2.7 \times 10^{-15}$ |
| Anisaldazine in water | 1.3 | 1.25 | 0.50 | $1.43 \times 10^{-12}$ |
| | | | 0.25 | $2.45 \times 10^{-13}$ |
| | | | 0.05 | $2.66 \times 10^{-15}$ |
| Anisilidenbenzidine in water | 0.81 | 1.25 | 0.50 | $8.95 \times 10^{-13}$ |
| | | | 0.25 | $1.65 \times 10^{-13}$ |
| | | | 0.05 | $2.05 \times 10^{-15}$ |

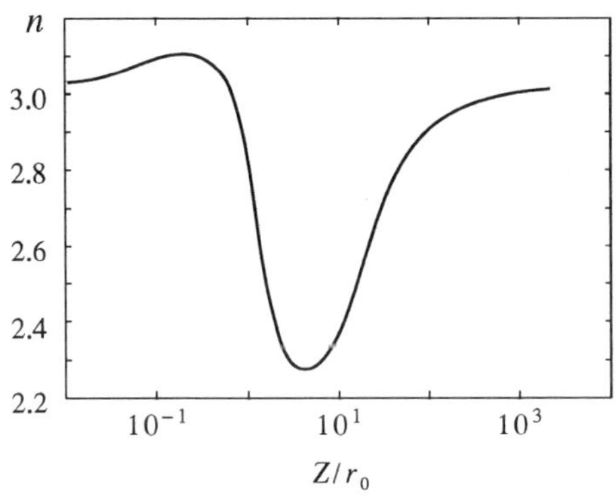


**Figure 4.2.** The plot of power index $n$ versus $Z/r_0$ ($p = 1.25$).

The dependence of $\gamma$ on $r$ is defined by (3.43) and (3.44). Numerical analysis of these expressions shows that if $r$ differs from arbitrary $r_0$ by less than a factor of 5, we can use a simpler relation for $\gamma$ and $r$: $\gamma \sim r^n$.

It will approximate (3.43) and (3.44) at any $Z$ and in a wide range of spheroid axis ratios $1.1 \leq p \leq 10$. The value of the power $n$ varies only with $Z/r_0$. A typical dependence of $n$ on $Z/r_0$ is shown in Figure 4.2 as calculated at $p = 1.25$.

One can see that $n \approx 3$ at fairly large and fairly small $Z/r_0$, which means that $\gamma$ is proportional to the particle volume ($\gamma \sim r^3$) at constant $Z$ ($Z$ in this case is practically the same for all particles of a system). When $Z/r_0 = 4.5$, the value of $n$ is minimum and equal 2.25, i.e. $n \approx 2$. Then $\gamma$ is proportional to the surface area of a particle ($\gamma \sim r^2$). Such a dependence was experimentally observed for diamond, anisaldazine and anisilidenbenzidine. Figure 4.3 shows the experimental plots of $N(E^2)$ for the above colloidal dispersions together with the curves calculated on the assumption that $\gamma \sim r^2$ (solid curves) and $\gamma \sim r^3$ (dashed curves). As seen from the Figure, the experimental plots for all of the dispersions are best described by the dependences calculated

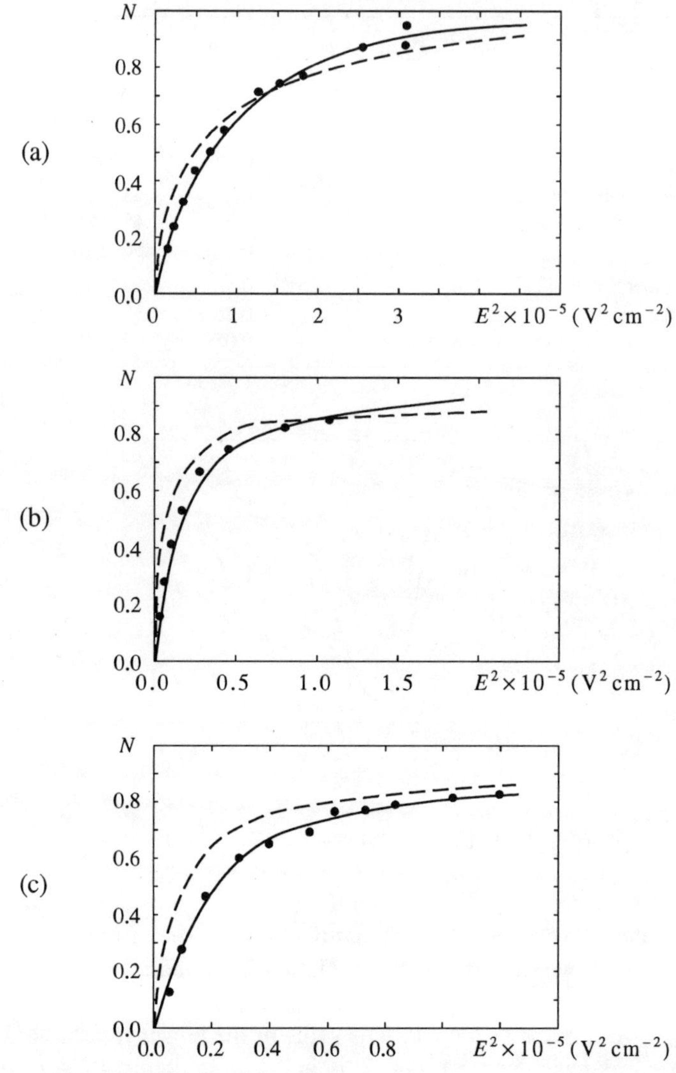


**Figure 4.3.** The plots of $N(E^2)$ (●, experiment; solid curves, theory on the assumption that $\gamma \sim r^2$; dashed curves, theory at $\gamma \sim r^3$) for different dispersed materials in water. (a) Diamond. (b) Anisaldazine. (c) Anisilidenbenzidine.

on the assumption that $\gamma \sim r^2$. It should be noted that the Dukhin–Shilov theory allows the conditions for which $2 < n < 3$, while in the Maxwell–Wagner theory $n$ is always equal to 3, which disagrees with the experimental $E^2$ dependences of $N$ in Figure 4.3.

It would be also interesting to compare the polarizability data obtained from a field-induced dichroism study with those obtained by other electrooptic

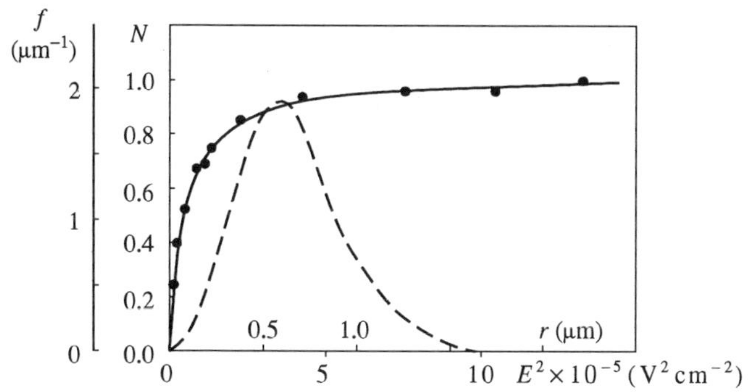


**Figure 4.4.** The plots of $N(E^2)$ (●, experiment; solid curve, theory) and $f(r)$ (dashed curve) for a benzopurpurine colloidal suspension.

methods. For such a comparison, we have selected a benzopurpurine aqueous colloidal dispersion because it was studied in detail by the Bulgarian research group.[59] The authors presented measurements of the relative variation in the scattered intensity $\alpha_{sc}$ as a function of the squared field. They also obtained the dependence of scattering anisotropy on the electric pulse duration and the relaxation behavior of anisotropy at zero field, i.e. in the interval between two successive pulses. These data have permitted a comparison of the method used by the Bulgarian researchers with our own approach. Here we describe our method and the results of the comparison.

In order to find the anisotropy of particle polarizability, we studied the experimental dependence of $N(E^2)$ associated with induced dichroism on the sine-shaped electric field square. The applied field frequency was 5 kHz. The measured $N(E^2)$ function is shown in Figure 4.4 (solid curve). Here one can also see the function $f(r)$ for the dichroic-weighted size distribution of particles (dashed curve) obtained from the relaxation time dependence of $N$ following the field switch-off (see Chapter 2). This function is necessary for description of induced dichroism $N$ in a polydisperse system (see (3.43), (3.44), (4.1), and (4.2)).

With the least square method, we have computed the $Z$ value at which the theoretical curve for $N(E^2)$ best fits the experimental dependence. The discrepancy between these curves is a criterion for the theory applicability. The theoretical $N(E^2)$ in Figure 4.4 is shown by a solid curve which fits well with the experimental values represented by dots. This confirms the applicability of the polarizability theory for spheroids to real benzopurpurine particles. The value of $Z = 10^{-4}$ cm is found at the point where the experimental and theoretical curves overlap.

Another theoretical parameter of interest is the semiaxis ratio $p$. We have found it from electron microscopic data to be 2.5. We will show later that the value of $Z$ is not very sensitive to slight variation in $p$ around the

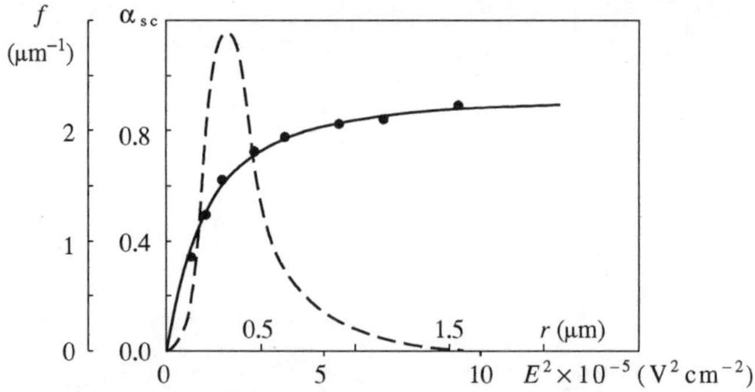


**Figure 4.5.** The plots of $\alpha_{sc}(E^2)$ (•, experiment; solid curve, theory) and $f(r)$ (dashed curve) for a benzopurpurine colloidal suspension.

above value. For example, in the range of $p$ between 2 and 5, $Z$ varies from $1.2 \times 10^{-4}$ to $0.9 \times 10^{-4}$ cm. This permits the solution of the polarizability problem for benzopurpurine particles in a polydisperse, though monomorphic approximation. The quantity $Z$ is determined only by the physical and chemical properties of the particle surface and of the dispersive medium. It does not depend on the particle size and is a major characteristic of particle polarizability in the polydisperse system considered.

The obtained value of $Z$ permits calculation of $\gamma$ for particles of specific size and shape. In the benzopurpurine colloidal system the $\gamma$ values for particles markedly contributing to $N$ lie between $2.4 \times 10^{-13}$ and $4.1 \times 10^{-11}$ $cm^3$ for the sizes $r$ from 0.15 to 1.5 $\mu$m (Figure 4.5). It is quite likely that these limits do not cover the entire range of $\gamma$ and $r$, and the colloidal system may also contain smaller particles with lower $\gamma$ and larger ones with higher $\gamma$, which may not possess dichroism and, therefore, make no contribution to the optical effect.

Let us now turn to the data obtained by the Bulgarian group.[59] They studied relative variations in the intensity of light scattered by a benzopurpurine colloidal solution when an electric field was applied to it (the $\alpha_{sc}$-effect). It can be shown that the expressions for the dependence of $\alpha_{sc}$ and $N$ on particle orientation in an electric field are identical. This circumstance permits us to apply the technique that was suggested for finding $N$ to the study of the $\alpha_{sc}$-effect. We have processed the $N(t)$ relaxation curve[59] and obtained a curve for the "optically weighted" size distribution due to light scattering (see Figure 4.5). The curve for $\alpha_{sc}(E^2)$ calculated at $Z = 1.5 \times 10^{-4}$ cm fits well with the experimental data. The particle shape again corresponds to $p = 2.5$, and the theory here is in good agreement with experiment.

It is noteworthy that the value of $Z$ obtained experimentally by the Bulgarian group differs only by 15% from that found in experiments on $N$. For such a value of $Z$ the anisotropy of polarizability $\gamma$ varies from $2.6 \times 10^{-13}$ to $2.7 \times 10^{-11}$ $cm^3$ for particle sizes $r$ in the range 0.15 to 1.2 $\mu$m. These are the particles that do contribute to the $\alpha_{sc}$-effect, as follows from the curve for the optically weighted size distribution.

The same workers[59] made a comparison between the values of electric polarizability $\gamma$ measured in three ways: from the $\alpha_{sc}(E^2)$ slope to the axis $E^2$ at $E = 0$; from the electric field $E$ minimum at which the $\alpha_{sc}$-effect was the strongest; from the time during which $\alpha_{sc}$ increased by 63% of its steady-state value after the external field was applied. The results obtained in these three ways differ nearly by a factor of 50. No doubt, this discrepancy results from the polydispersity of the colloidal system. At low fields, only large particles become oriented, because $\gamma$, and therefore the energy of their interaction with the field that determines the degree of orientational ordering, are much larger than for small particles.

When finding $\gamma$ from the field value $E$ at which the optic effect saturates, one should keep in mind that at high fields large particles become completely oriented along the field, so only small particles with low $\gamma$ can contribute to the variation of the electrooptic effect. When one derives $\gamma$ from the time of the electrooptic effect growth, one finds only an intermediate value. Our studies have shown this value to vary significantly with the electric field $E$ and to increase with decreasing $E$. Naturally, the values of $\gamma$ found in the three ways must be very close for a strictly monodisperse colloidal solution.

**Substantiation of monomorphic approximation.** In our calculations of polarizability anisotropy in colloidal solutions, we took into account the system polydispersity. The shape of particles was approximated by elongated spheroids and given by a mean value of $p$ found from electron microscopic observations. In the calculations $p$ was taken to be the same for all particles. With a few exceptions, $p$ is variable in real dispersions; for this reason, the monomorphic approximation for calculation of $\gamma$ and $Z$ is to be substantiated.

The way of finding the dichroic-weighted size distribution function $f(r)$ was described in Chapter 2. It was based on the study of orientation relaxation due to the particle rotational diffusion after the field turn-off. This method allows one to find $f(r_D)$, where $r_D$ is the radius of a sphere, whose rotational diffusion coefficient $D(r_D)$ is equal to that of a particle, $D$. The coefficient $D$ is defined not only by the size but also by the structure factor $F$:

$$D = \frac{kT}{6FV\eta}. \tag{4.3}$$

With the particle shape and $f(r_D)$, one can find the function $f(r)$ if one knows the correlation between $r$ and $r_D$ ($r$ is the radius of a sphere with a volume equal to that of the particle). We obtain this correlation from the equality $D(r_D) = D$:

$$r = \frac{r_D}{F^{1/3}}. \tag{4.4}$$

The analytical expression for $p$ dependences of $F$ for particles approximated by elongated ellipsoids of revolution is:[33]

$$F(p) = \frac{2(p^4-1)}{3p^2}\left(\frac{2p^2-1}{2p\sqrt{p^2-1}}\ln\frac{p+\sqrt{p^2-1}}{p-\sqrt{p^2-1}} - 1\right)^{-1}. \tag{4.5}$$

For strongly elongated particles, this expression can take the form

$$F(p) = \frac{p^2}{3\ln(2p-0.5)}.$$

Relations (3.43) and (3.44) for particle polarizabilities along the long and short axes, with account of (4.4), are:

$$\gamma_{\|} = \frac{\epsilon r_D^3}{3F}\left(\frac{3}{2}I_{\|}g - 1\right)\left[1 + A_x\left(\frac{3}{2}I_{\|}g - 1\right)\right]^{-1} \tag{4.6}$$

$$\gamma_{\perp} = \frac{\epsilon r_D^3}{3F}\left(\frac{3}{2}I_{\perp}g - 1\right)\left[1 + \frac{1-A_x}{2}\left(\frac{3}{2}I_{\perp}g - 1\right)\right]^{-1}, \tag{4.7}$$

where $g = Z(pF)^{1/3}/r_D$.

They define the function of polarizability anisotropy $\gamma(p)$ at a given $Z$, if $D(r_D)$ is constant, i.e., $r_D$ = const. The parameters that define this function are $p$, $F$, $I_{\|}$, $I_{\perp}$ and $A_x$.

A plot of $\gamma(p)$ for different $r_D$ calculated by formulas (4.6) and (4.7) is shown in Figure 4.6. One can see that a strong dependence of $\gamma$ on $p$ falls only in the range $1 < p < 2$. When $p$ varies from 2 to 20, the value of $\gamma$ changes less than by a factor of 1.5. It should be noted that the anisotropy of particle polarizability is very sensitive to this variation in $r_D$. For example, at $p = 2$ and $Z = 10^{-4}$ cm, the change in $r_D$ from $10^{-5}$ to $10^{-4}$ cm increases $\gamma$ by as much as a factor of 342 (see Figure 4.6b). These facts evidence that $\gamma$ of particles with $p \geq 2$ is strongly dependent on $r_D$ and weakly on the particle shape. The weak $p$ dependence of $\gamma$ indicates that the parameter $Z$ for such particles must also depend on $p$, when the form of the function $f(r_D)$ remains the same. Let us find this dependence.

The values of $Z$ will be found, as previously, from the comparison between experimental and theoretical $N(E^2)$. If the function $f(r_D)$ is given, formula (4.1) for the function $N(E^2)$ can be written as follows:

$$N = \int_0^{\infty} f(r_D)\mathcal{F}(\gamma, E)\mathrm{d}r_D. \tag{4.8}$$

The form of the function $\mathcal{F}(\gamma, E)$ is defined by (4.2), while $\gamma_{\|}$ and $\gamma_{\perp}$ are described by (4.6) and (4.7). Retaining the form of $f(r_D)$, we can find the $p$ dependence of $Z$ by employing the above procedure.

Figure 4.7a shows the function $Z(p)$ for a benzopurpurine colloidal system together with the function $f(r_D)$ necessary for finding $Z$. One can see that $Z$ varies from $1.6 \times 10^{-4}$ to $8.8 \times 10^{-5}$ cm when $p$ ranges between 1.7 and 20.

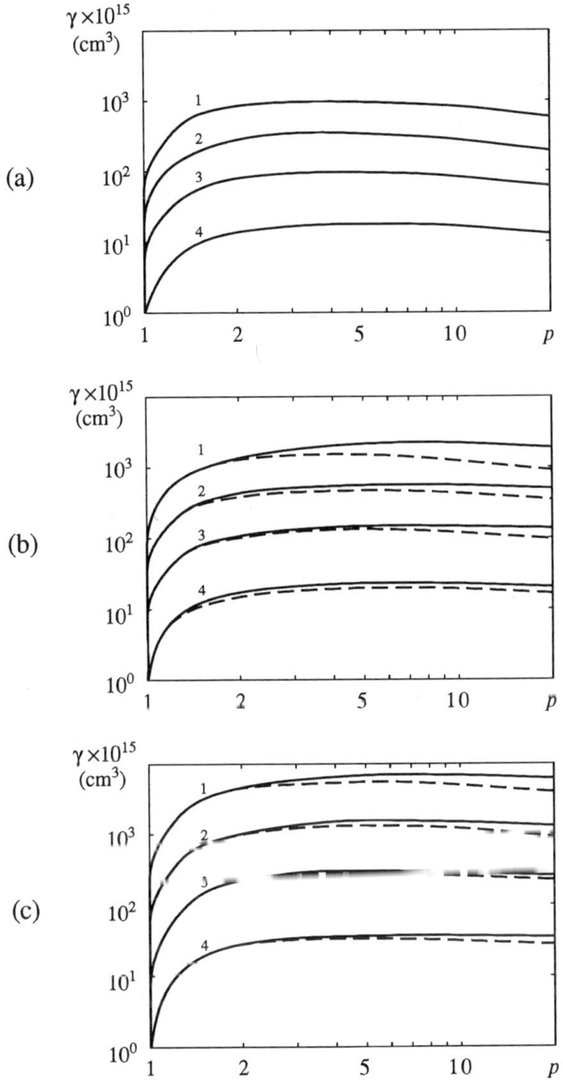


**Figure 4.6.** The plots of $\gamma$ for different values of $r_D$: $1.0 \times 10^{-4}$ cm (curves 1), $5.0 \times 10^{-5}$ (curves 2), $2.5 \times 10^{-5}$ (curves 3), $1.0 \times 10^{-5}$ (curves 4). Elongated spheroids (solid curves), flattened spheroids (dashed curves). (a) $Z = 5.0 \times 10^{-5}$ cm, (b) $Z = 1.0 \times 10^{-4}$, (c) $Z = 2.0 \times 10^{-4}$.

The dependence $N(E^2)$ calculated for different $p$, as well as the experimental $N(E^2)$, are plotted in Figure 4.7b. The calculations show that the shapes of the curves do not, in fact, depend on $p$ if it varies from 1.7 to 20 (one should certainly keep in mind that the values of $Z$ for these curves are different). Only at $p = 1.5$ the difference between the calculated and experimental $N(E^2)$

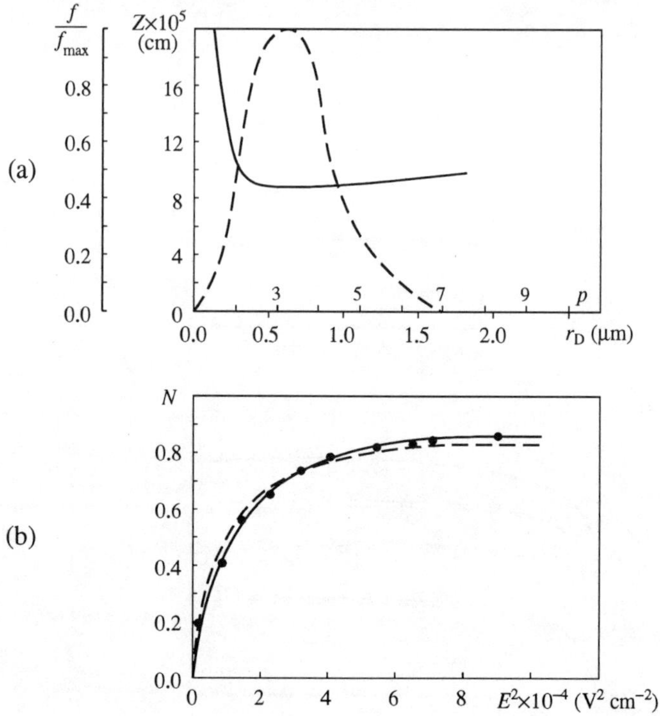


**Figure 4.7.** A benzopurpurine colloidal suspension. (a) The plots of $Z(p)$ (solid curve) and $f(r_D)$ (dashed curve). (b) The plots of $N(E^2)$. Solid curve, theory at $p = 1.5$; dashed curve, theory at $p = 2.5$; •, experiment.

exceeds an experimental error. It should be noted that the change of $r_D$, (i.e., the scale of the function $f(r_D)$) only by a factor of 1.5 would change $Z$ from $1 \times 10^{-4}$ to $6.2 \times 10^{-4}$ cm at $p = 2.5$.

The values obtained in this way are highly sensitive to the form of the function $f(r_D)$, being only weakly dependent on the particle shape at $p > 2$. If we take into account the particle shape distribution and employ the distribution function with respect to the semiaxis ratio $p$ to determine $Z$, then $Z$ will depend more weakly on the mean $p$ than it does in the monomorphic approximation. For example, a uniform distribution of particle shapes at $p$ between 1.7 and 3.5 for a benzopurpurine colloidal dispersion would give $Z = 1.1 \times 10^{-4}$ cm, which differs only by 10% from the value derived in the monomorphic approximation at $p = 2.5$. This approximation does not much distorts the true value of $Z$ in the above method for dispersions with sufficiently elongated particles ($p \geq 2$).

If the particles are not so much elongated ($p < 2$), the $p$ dependence of $Z$ is quite different. In Figure 4.6 the value of $\gamma$ in this region of $p$ variation strongly depends on the particle shape when $r_D$ = const. The shape variation must, therefore, lead to a variation in $Z$.

A plot of $Z(p)$ for synthetic diamond in water is shown in Figure 4.8a.

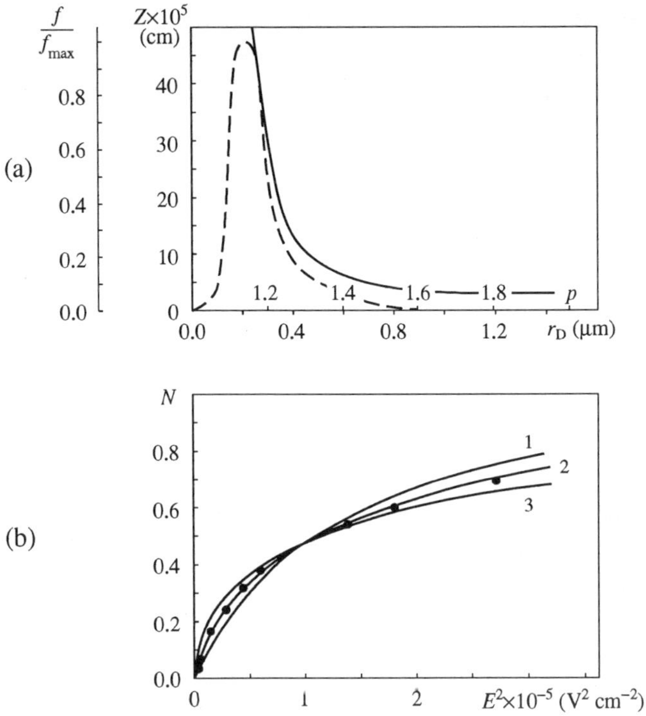


**Figure 4.8.** A a synthetic diamond suspension. (a) The plots of $Z(p)$ (solid curve) and $f(r_D)$ (dashed curve). (b) The plots of $N(E^2)$. Curve 1, theory at $p = 2.0$; curve 2, theory at $p = 1.25$; curve 3, theory at $p = 1.15$; •, experiment.

The change in $p$ from 1.15 to 2 results in a variation of $Z$ from $5.1 \times 10^{-4}$ to $3.0 \times 10^{-5}$ cm. Note that the theoretical function $N(E^2)$, unlike the one for sufficiently elongated particles, undergoes an appreciable variation with $p$ (Figure 4.8b). The best fit between the theoretical and experimental curves is observed at $p = 1.25$, which, as was pointed out above, is consistent with electron microscopic observations. The value of $Z$ in this case becomes equal to $1.6 \times 10^{-4}$ cm. The calculation shows that if the particle shape is taken to be uniformly distributed over the $p$ range from 1.2 to 1.4, then $Z$ is $1.45 \times 10^{-4}$ cm. Although this value is close to that derived from the monomorphic approximation for $p = 1.25$, the error in $Z$ due to the uncertainty of shape of slightly elongated particles ($p < 2$) may be fairly large and will increase with smaller elongation.

For irregular particles with a slight elongation, the parameter $p$ for a real particle is assumed to be equal to that for a spheroid, which would give the same $\gamma$ at the same $D$ as a real particle. Therefore, in order to determine $Z$ for slightly elongated particles, the theoretical and experimental functions $N(E^2)$ should be compared in two parameters: $Z$ and $p$. In this case the measurement error will strongly depend on the error in experimental $N(E^2)$, otherwise one should rigorously find the semiaxis ratio using another technique.

For a polymorphic dispersion containing irregular, slightly elongated particles, some particles can be well approximated by flattened rather than elongated spheroids. In order to substantiate the monomorphic approximation for finding $Z$, we must estimate how it changes in the transfer from the approximation by elongated spheroids to that by flattened ones.

The polarizability of flattened spheroids along the short and long axes can be represented as[60]

$$\gamma_{\|} = \frac{\epsilon r_D^3}{3F'}\left(\frac{3}{2}I'_{\|}g' - 1\right)\left[1 + \frac{1 - A'_x}{2}\left(\frac{3}{2}I'_{\|}g' - 1\right)\right]^{-1} \tag{4.9}$$

$$\gamma_{\perp} = \frac{\epsilon r_D^3}{3F'}\left(\frac{3}{2}I'_{\perp}g' - 1\right)\left[1 + A'_x\left(\frac{3}{2}I'_{\perp}g' - 1\right)\right]^{-1}, \tag{4.10}$$

where $A_x(p) = A'_x(p')$, $g' = Zp^{2/3}(F')^{1/3}/r_D$, $p' = 1/p$; $F'$ is the structure factor of a flattened spheroid; $I'_{\perp}$ and $I'_{\|}$ are the coefficients dependent on $p$. Following the work,[60] we have

$$I'_{\perp} = \frac{p'^2}{\sqrt{1 - p'^2}}\left(\frac{2 - p'^2}{1 - p'^2}\ln\frac{1 + \sqrt{1 - p'^2}}{p'} - \frac{1}{\sqrt{1 - p'^2}}\right)$$

$$I'_{\|} = \frac{p'^2}{2}\left(\frac{2 - p'^2}{(1 - p'^2)^{3/2}} - \frac{p'^4}{1 - p'^2}\ln\frac{1 + \sqrt{1 - p'^2}}{p'}\right).$$

The $p$ dependence of $A_x$ follows from (3.21). The analytical expression for $F'(p)$ has the form[33]

$$F'(p) = \frac{2}{3}\frac{1 - p'^4}{p'^2}\left(1 + \frac{1 - 2p'^2}{p'\sqrt{1 - p'^2}}\arctan\frac{\sqrt{1 - p'^2}}{p'}\right)^{-1}.$$

Relations (4.9) and (4.10) describe the $p$ dependence of the polarizability anisotropy $\gamma$ at a given $Z$ if $D(r_D)$ remains constant, i.e., $r_D$ = const. Figure 4.6b,c gives the functions $\gamma(p)$ at different $r_D$ calculated from (4.9) and (4.10) for the $Z$ values of $1.0 \times 10^{-4}$ and $2.0 \times 10^{-4}$ cm, respectively. One can see that in the range of $p$ from 1 to 2 the $\gamma$ values for flattened spheroids are close to those calculated for elongated spheroids. Therefore, the transfer to the flattened particle approximation is unlikely to lead to noticeable variations of this parameter.

Figure 4.9 shows the function $Z(p)$ for synthetic diamond in an aqueous solution calculated in the flattened spheroid approximation. It is close to $Z(p)$ determined in the elongated spheroid approximation (see Figure 4.8). Figures 4.8b and 4.9 show that the shape of the $N(E^2)$ curve essentially

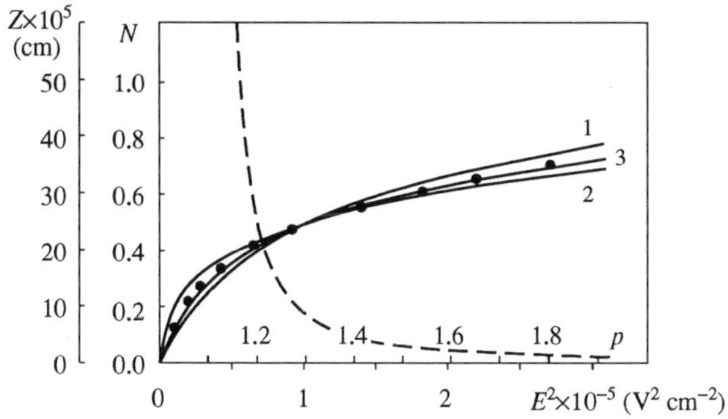


**Figure 4.9.** A diamond suspension. The plots of $Z(p)$ (dashed curve) and $N(E^2)$ (solid curves). Curve 1, theory at $p = 1.7$; curve 2, theory at $p = 1.25$; curve 3, theory at $p = 1.15$; •, experiment.

changes with varying $p$. The best fit between the theoretical and experimental dependences is at $p = 1.25$. Therefore, the transfer from the approximation by elongated spheroids to that by flattened spheroids does not change appreciably the parameter $Z$ for slightly elongated particles.

To conclude, the analysis of monomorphic approximation for finding $Z$ shows that this approximation does not lead to a significant error, if the system under study contains sufficiently elongated particles ($p \geq 2$). If, however, the particles are only slightly elongated ($p < 2$), the uncertainty in the particle shapes may result in an essential error up to 100%.

**Study of aqueous palygorskit colloid.** Palygorskit is a clay which has become an interesting object for investigation. Electron microscopic studies of this mineral[61] have shown that a palygorskit particle represents a strongly elongated rod with an average length of $10^{-4}$ cm and a thickness of $4.0 \times 10^{-6}$ cm ($p = 25$). The physical and chemical properties of a palygorskit aqueous colloidal dispersion are understood quite well.[58,61] Some workers[62,63] have presented electrooptic data on light scattering anisotropy induced by an external sine-shaped field. So it is tempting to compare these data with the results of our electrooptic studies of this dispersion.

A study of electrooptic characteristics of a palygorskit colloidal dispersion meets with certain experimental difficulties, because the particle volume is relatively small, $10^{-15}$ cm$^3$, i.e., $10^2$–$10^3$ times less than that of diamond, anisaldazine, benzopurpurine and other particles we have investigated. The small particle volume leads to weak electrooptic effects. Higher number densities of dispersed material increase the amplitudes of electrooptic effects, however, the density can be increased only up to a certain limit beyond which the particles begin to aggregate.

Formation of aggregates increases the relaxation time of an electrooptic effect after the external field is switched off.

Figure 4.10 shows the values of $N$ as a function of time, $N(t)$, after the

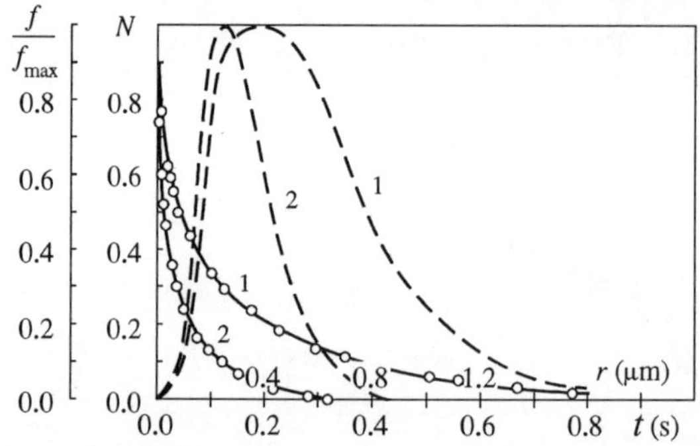


**Figure 4.10.** The plots of $N(t)$ (solid curves) and $f(r)$ (dashed curves) for a palygorskit suspension with the number densities: $c = 10^{10}$ cm$^{-3}$ (curves 1). $c = 10^7$ cm$^{-3}$ (curves 2).

field switch-off for a palygorskit colloidal dispersion with the number densities of $10^7$ cm$^{-3}$ (for the mean interparticle spacing $4.6 \times 10^{-3}$ cm) and $10^{10}$ cm$^{-3}$ (the mean spacing $4.6 \times 10^{-4}$ cm). The same Figure gives the dichroic-weighted size distribution functions $f(r_D)$ obtained from the $N(t)$ curves. It is seen that the relaxation times of the effect differ by nearly an order of magnitude and the mean effective sizes differ by a factor of two for colloidal solutions with the above number densities. This fact evidences the formation of aggregates in a palygorskit system with the number density $10^{10}$ cm$^{-3}$. Note that the shape of the relaxation curve for $N(t)$ for concentrations less than $10^7$ cm$^{-3}$ remains the same.

All the subsequent studies of electrooptic properties of this colloidal dispersion were carried on at the number density $10^7$ cm$^{-3}$. Prior to measurements, the systems were always subjected to ultrasonic treatment at a frequency of 21 kHz to prevent the formation of particle aggregates.

Electron microscopic studies of palygorskit particles[61] show that their shape varies in a wide range ($p$ ranges from 5 to 100) together with the thickness and length. To determine the parameter $Z$, one should know the shape and size distributions described by the function $f_1(r, p)$. If the shape distribution does not depend on the size distribution, the function $f_1(r, p)$ can be represented as the product $f(r)\, f_2(p)$, where $f(r)$ describes the size distribution and $f_2(p)$ the shape distribution of particles. Unfortunately, we had no way of finding the function $f_2(p)$ and had to use the following two approximations in order to find $p$.

(i) All the particles were assumed to have the same shape characterized by a mean $p$ value of 25 determined from electron microscopic observations. The size distribution in this case was described by the function $f(r)$.

(ii) The particle thickness was assumed to be equal to a mean value of $2b = 4 \times 10^{-6}$ cm. The size distribution was described by the function $f(L)$, where $L$ is the particle length. To find $f(L)$ and $f(r)$, we used the function $f(r_D)$.

The relationship between $r_D$ and $r$ in the first case was found from (4.4) and in the second case the relationship between $L$ and $r_0$ can be obtained by equating the rotational diffusion coefficient for a cylinder of length $L$ to that of a sphere:

$$\frac{3kT}{\pi\eta L^3}(\ln 2p - 0.5) = \frac{kT}{8\pi\eta r_D}. \tag{4.11}$$

The function $N(E^2)$, if $p = \text{const}$, is described by expressions (4.1), (3.42), and (3.43). If $b = \text{const}$, it can be written as

$$N = \int_0^\infty f(L)\mathcal{F}(\gamma, E)\mathrm{d}L.$$

The function $\gamma(L) = \gamma_\parallel(L) - \gamma_\perp(L)$ is derived from (4.6), (4.7), and (4.11):

$$\gamma_\parallel = \epsilon\frac{b^2L}{3}\left(\frac{3}{2}I_\parallel\frac{Z}{b} - 1\right)\left[1 + A_x\left(\frac{3}{2}I_\parallel\frac{Z}{b} - 1\right)\right]^{-1} \tag{4.12}$$

$$\gamma_\perp = \epsilon\frac{b^2L}{3}\left(\frac{3}{2}I_\perp\frac{Z}{b} - 1\right)\left[1 + \frac{1 - A_x}{2}\left(\frac{3}{2}I_\perp\frac{Z}{b} - 1\right)\right]^{-1}. \tag{4.13}$$

By repeating the above procedure for deducing $Z$ from the comparison of the theoretical and experimental functions $N(E^2)$, we obtain $Z = 2.3 \times 10^{-4}$ cm at $p = \text{const}$ and $Z = 2.1 \times 10^{-4}$ cm at $b = \text{const}$.

Figure 4.11a gives an experimental plot of $N(E^2)$ for a palygorskit aqueous colloidal system and a theoretical $N(E^2)$ obtained at $Z = 2.3 \times 10^{-4}$ cm and $p = 25$. It also gives the function $f(r)$ necessary for the calculation of $N(E^2)$. Figure 4.11b shows the same dependences calculated at $Z = 2.1 \times 10^{-4}$ cm and $b = 2.0 \times 10^{-6}$ cm. Though the values of the parameters are close to each other, the theoretical curve for $N(E^2)$ in the first case is somewhat closer to the experimental one. This indicates that $f(L)$ provides a better description of the particle size distribution in a real palygorskit dispersion than $f(r)$.

With $Z$ known, it is possible to determine the anisotropy of polarizability $\gamma$ if particle sizes and shapes are known too. A plot of $\gamma$ versus length $L$ calculated from (4.12) and (4.13) is presented in Figure 4.12. The particle thickness was assumed to be constant and equal to $4.0 \times 10^{-6}$ cm, and $Z$ was taken to be $2.1 \times 10^{-4}$ cm. The light scattering technique[62] was used to study palygorskit aqueous systems with smaller particles, $L = 0.34$ $\mu$m. The calculated value of $\gamma$ was $3.8 \times 10^{-13}$ cm$^3$. It strongly differs from the value $5.5 \times 10^{-14}$ cm$^3$ we have obtained in studies of field-induced dichroism in palygorskit dispersions, and this difference is very likely to be due to a monodisperse approximation.[62]

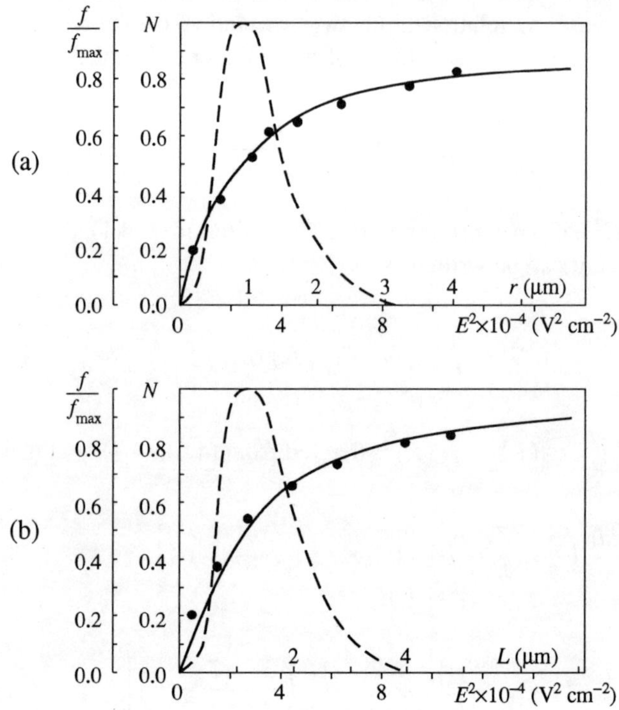


**Figure 4.11.** A palygorskit aqueous colloidal suspension. The plots of $N(E^2)$ (•, experiment; solid curves, theory), $f(r)$ and $f(L)$ (dashed curves). (a) For the spheroidal approximation. (b) For the cylindrical approximation.

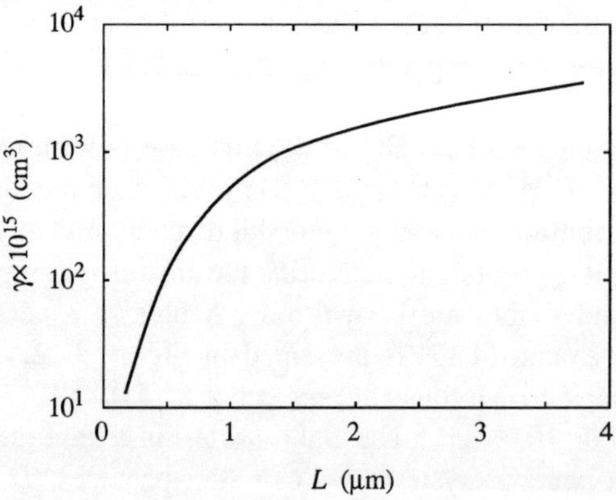


**Figure 4.12.** A plot of $\gamma$ calculated at $Z = 2.1 \times 10^{-4}$ cm, and $b = 2 \times 10^{-6}$ cm.

## 4.3. EFFECT OF THE MEDIUM IONIC COMPOSITION ON POLARIZABILITY ANISOTROPY

According to (3.43) and (3.44), the anisotropy of polarizability $\gamma$ for colloidal particles is extremely $Z$-dependent. The value of $Z$ can be varied by varying the ionic composition or concentration of the fluid, which entails variation in $\kappa^\sigma$ and in the fluid electrical conductivity $K$, both of which define $Z$. The value of $N$ associated with particle orientation in an external field directly depends on $\gamma$ and, hence, it must also vary with the fluid ionic composition. For example, the electrooptic effect $N$ arising from the dichroism of colloidal particles will weakly depend on the field $E$ as $\gamma$ decreases. Our objective was to follow variations in $\gamma$ and $\kappa^\sigma$ with varying ionic composition of the fluid, using the experimental measurements of $N$. The study was done in a synthetic diamond aqueous colloidal suspension with $Th(NO_3)_4$ and $AlBr_3$ electrolyte additions and in a palygorskit aqueous colloidal system with NaCl electrolyte additions.

If the electrical conductivity of a fluid is too high or the polarizability anisotropy is too low, then at electric fields high enough to produce considerable particle orientation, the dispersion becomes heated, which sometimes makes the study impossible. To determine $Z$ in such colloidal systems, one can restrict the study to fields inducing slight particle orientation. The values of $I^0_\parallel$ and $I^0_\perp$ then should be determined in systems with the same number density but in a different dispersive medium, which would have the same refractive index and be capable of providing a complete orientation. For slight orientations, the function $\mathcal{F}(\gamma, E)$ defined by (4.2) may have the form:

$$\mathcal{F}(\gamma, E) = \frac{\gamma E^2}{15kT}, \tag{4.14}$$

then (4.1) describing the dependence of $N$ on $E$ takes the form:

$$N = \frac{E^2}{15kT} \int_0^\infty f(r)\gamma \mathrm{d}r. \tag{4.15}$$

This expression can be used to find $Z$ from a comparison of theoretical and experimental values of $N$. It should be stressed that (4.15), in contrast to (4.8), cannot provide unambiguously the parameter $Z$ for the given dispersion, since the polarizability anisotropy is not a monotonic function of $Z$ (see Figure 4.3). It can be used, however, to find a relative variation in $Z$, as the physical and chemical properties of the colloidal suspension change.

Let us introduce the magnitude $\beta$ defined by the slope of the $N(E^2)$ curve at $E = 0$:

$$\beta = \left. \frac{\mathrm{d}N}{\mathrm{d}E^2} \right|_{E=0}.$$

In the field range producing a slight particle orientation, $\beta$ is related to the polarizability anisotropy $\gamma$ as:

$$\beta = \frac{1}{15kT} \int_0^\infty \gamma f(r) \mathrm{d}r. \tag{4.16}$$

This relation allows us to follow variation in $\gamma$, even at low $\gamma$ and high $K$, from the changes in $\beta$ determined experimentally.

We have found correlations between $N$ and $E$ for all the dispersions studied. The selected field frequency of 1.5 kHz corresponded to a frequency range, in which $N$ was frequency-independent.[64] Analysis of the $N(E^2)$ curves could yield the value of $\beta$. In the systems with $Th(NO_3)_4$, $AlBr_3$ and NaCl additions, the anisotropy of particle polarizability varied in a wide range, and in some cases we failed to get a total orientational ordering necessary to determine $N$ and $\beta$. For this reason, all of the colloidal suspensions we studied had the same number density of $10^6$ cm$^{-3}$ for diamond and $10^7$ cm$^{-3}$ for palygorskit.

One can see from relation (4.16) that variation in $f(r)$ can essentially change $\beta$. Therefore, a comparison of $\beta$ values for different systems is justified only when $f(r)$ is the same. Indeed, as pair aggregates are formed, they become oriented, producing an optic effect in lower fields than those necessary for inducing a single-particle electrooptic effect. This in turn increases the value of $\beta$.

At the number density selected for the experiment, the interparticle spacings were an order of magnitude larger than the particle sizes. At such a low density, there should be no appreciable coagulation even at the isoelectric point. Besides, all the colloidal suspensions had been treated with ultrasound to prevent coagulation. The medium composition was controlled with the $N(t)$ curves indicating the time variation of the optical signal following the field turn-off. Formation of pair aggregates would decrease the rotational diffusion constants by a few factors as compared with single particles (see (4.3)), which would appreciably change the shape of the $N(t)$ curve.

Figure 4.13 gives the $N(t)$ dependences for diamond suspensions at different $Th(NO_3)_4$ concentrations. In order to evaluate $N(t)$, we selected fields capable of inducing the same values of $N$. Identical $N$ values indicate an identical degree of particle ordering if, of course, the size distribution is constant. Identical relaxation behavior of $N$ evidences for an identical composition of the systems under study. It is clear from Figure 4.13 that the $N(t)$ dependences nearly overlap, indicating the absence of aggregates in the dispersion. Similar results have been obtained for diamond suspensions with $AlBr_3$ additions and for palygorskit systems with NaCl additions.

The experimental dependences of $\beta$ on $Th(NO_3)_4$ and $AlBr_3$ electrolyte concentrations $c$ for diamond suspensions are shown in Figure 4.14. One can see that all the curves have a minimum. The non-monotonic character of the $\beta(c)$ curves can be interpreted in terms of the theory of polarizability anisotropy for spheroids.[51,60,65] Higher electrolyte concentration leads to a higher electrical

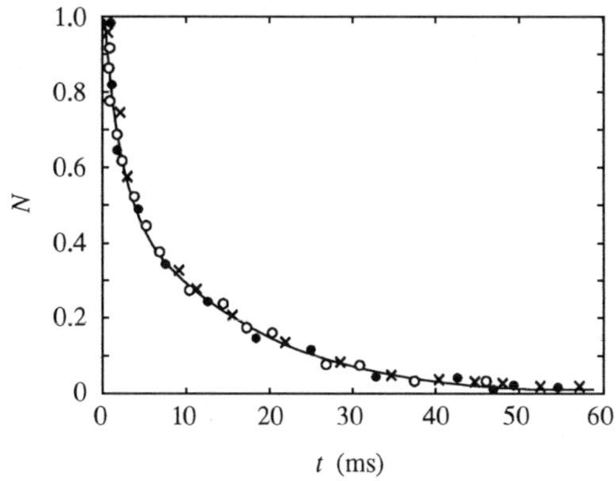


**Figure 4.13.** A diamond aqueous suspension with added $Th(NO_3)_4$ electrolyte. The plots of $N$ at different electrolyte concentrations. $\times, c = 1.2 \times 10^{-6}$ mol dm$^{-3}$; $\bullet, c = 1.7 \times 10^{-6}$ mol dm$^{-3}$; $\bigcirc, c = 5.3 \times 10^{-6}$ mol dm$^{-3}$.

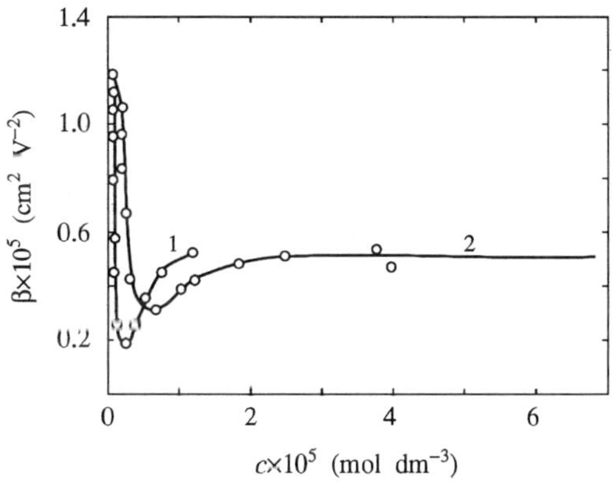


**Figure 4.14.** A diamond aqueous suspensions with an added electrolyte. The plots of $\beta$ versus electrolyte concentration. $Th(NO_3)_4$ (curve 1) and $AlBr_3$ (curve 2).

conductivity of the medium $K$ and to a lower specific surface conductivity of the colloidal particles and, hence, of $Z = \kappa^\sigma/K$. Moreover, $Z$ will decrease with increasing valency of counterions. For the diamond suspension (see Section 4.1), $Z = 1.6 \times 10^{-4}$ cm, so on addition of the electrolyte, $Z$ must become lower than this value.

A plot of $\beta(Z)$ calculated for a diamond colloidal solution by formulas (3.43) and (3.44) is shown in Figure 4.15 together with the $f(r)$ curve necessary for such calculations. The curve of $\beta(Z)$ also has a minimum at $Z = 1.5 \times 10^{-5}$ cm. It follows from a comparison of $\beta(c)$ and $\beta(Z)$ represented in Figures 4.14 and 4.15, respectively, that the concentrations of $Th(NO_3)_4$ and $AlBr_3$, responsible for the $\beta(c)$ minima and equal to $2.0 \times 10^{-6}$

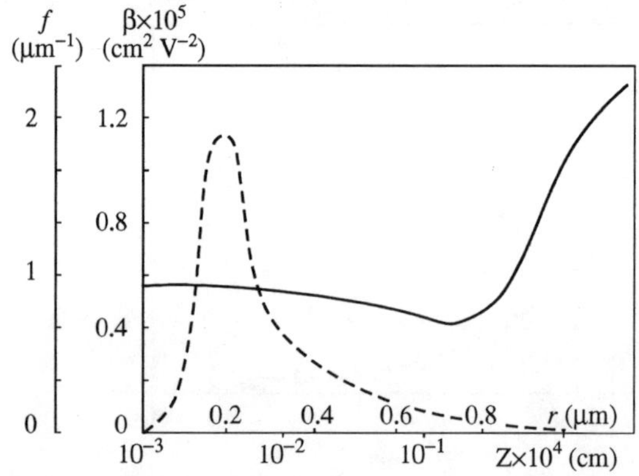


**Figure 4.15.** The plots of $\beta$ (*Z*) (solid curve) and $f(r)$ (dashed curve) for diamond particles in water.

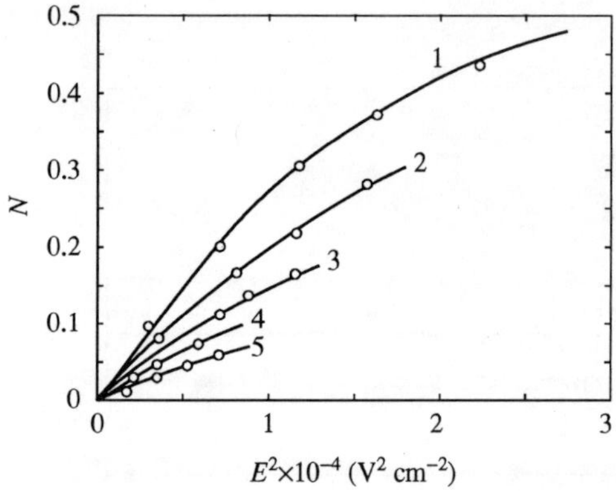


**Figure 4.16.** The plots of *N* for a palygorskit aqueous colloidal suspension at various NaCl electrolyte concentrations. $c = 1.0 \times 10^{-4}$ mol dm$^{-3}$ (curve 1); $c = 2.0 \times 10^{-4}$ mol dm$^{-3}$ (curve 2); $c = 3.0 \times 10^{-4}$ mol dm$^{-3}$ (curve 3); $c = 4.0 \times 10^{-4}$ mol dm$^{-3}$ (curve 4); $c = 6.0 \times 10^{-4}$ mol dm$^{-3}$ (curve 5).

and $6.0 \times 10^{-6}$ mol dm$^{-3}$, correspond to $Z = 1.5 \times 10^{-5}$ cm. The electrical conductivities of the dispersions corresponding to these concentrations of $Th(NO_3)_4$ and $AlBr_3$ are $5.0 \times 10^{-6}$ and $6.0 \times 10^{-6}\ \Omega^{-1}\,\mathrm{cm}^{-1}$, respectively. The conductivities were determined using a bridge circuit at a frequency of 1.5 kHz. With $Z$ and $K$ known, we were able to find the surface conductivities of colloidal particles which, at the concentrations of the $\beta(c)$ minima, turned out to be $7.5 \times 10^{-11}$ and $9.0 \times 10^{-11}\ \Omega^{-1}$ for the diamond suspensions with $Th(NO_3)_4$ and $AlBr_3$, respectively. The surface conductivity of diamond particles in solutions without ion admixtures was $1.6 \times 10^{-10}\,\Omega^{-1}$ ($K = 1.0 \times 10^{-6}\,\Omega^{-1}\,\mathrm{cm}^{-1}$, $Z = 1.6 \times 10^{-4}$ cm). Consequently, when $Th(NO_3)_4$ concentration varied from 0 to $2.0 \times 10^{-6}$ mol dm$^{-3}$ and $AlBr_3$

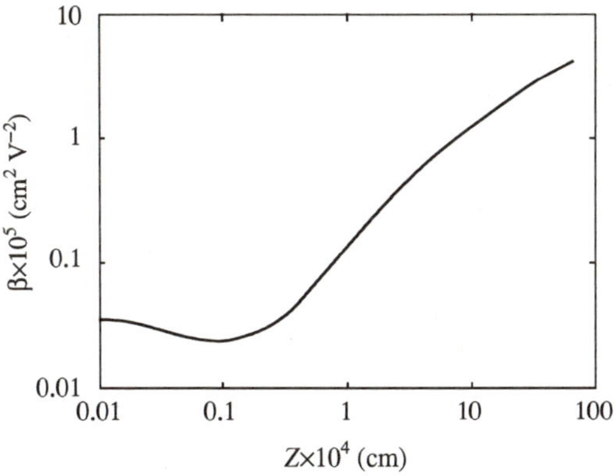


**Figure 4.17.** A plot of $\beta$ for a palygorskit aqueous colloidal suspension.

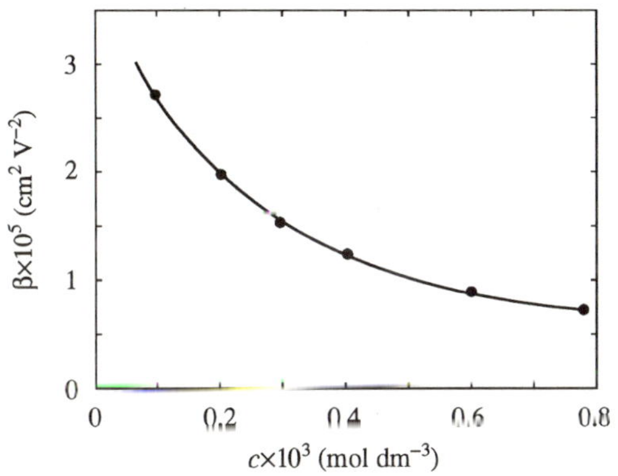


**Figure 4.18.** A plot of $\beta$ for a palygorskit aqueous colloidal suspension with NaCl electrolyte.

concentration from 0 to $6.0 \times 10^{-6}$ mol dm$^{-3}$, the conductivity $K$ of the systems increases by a factor of 5 and 6, whereas $\kappa^{\sigma}$ decreases by a factor of 1.8 and 2.1, respectively. Similar variations in $K$ and $\kappa^{\sigma}$ with increasing electrolyte concentration in palygorskit colloidal solutions were observed.[58]

The $Th(NO_3)_4$ and $AlBr_3$ electrolyte concentrations at the isoelectric point were $1.3 \times 10^{-6}$ and $4.5 \times 10^{-6}$ mol dm$^{-3}$, respectively, and about 1.5 times lower than the electrolyte concentrations at the $\beta(c)$ minimum, i.e. the minimum mean anisotropy of polarizability $\gamma$. These data on the effect of medium ionic composition on the polarizability anisotropy of colloidal particles are consistent with modern theory of polarizability for spheroids.[47]

The dependence of $\kappa^{\sigma}$ on NaCl concentration was found from experiments[58] and from the expression for the contribution of specific surface conductivity $\kappa^{\sigma}$ to the conductivity of the dispersion. Table 4.2 lists the calculation results.

**Table 4.2**

| Data[39] | | | | |
|---|---|---|---|---|
| $c \times 10^3$, mol dm$^{-3}$ | $K \times 10^4$ $\Omega^{-1}$cm$^{-1}$ | $\kappa a$ | Rel | $\kappa^\sigma \times 10^9$, $\Omega^{-1}$ |
| 0.10 | 0.20 | 0.80 | 142 | 5.7 |
| 0.77 | 1.0 | 1.9 | 28.6 | 5.7 |
| 3.9 | 4.7 | 4.0 | 1.72 | 1.6 |
| Present data | | | | |
| $c \times 10^3$, mol dm$^{-3}$ | $K \times 10^4$ $\Omega^{-1}$ cm$^{-1}$ | $Z \times 10^5$, cm | | $\kappa^\sigma \times 10^9$ $\Omega^{-1}$ |
| 0.1 | 0.20 | 20 | | 4.0 |
| 0.2 | 0.31 | 13 | | 4.0 |
| 0.3 | 0.43 | 9.0 | | 3.9 |
| 0.4 | 0.54 | 7.0 | | 3.8 |
| 0.6 | 0.77 | 4.7 | | 3.6 |
| 0.8 | 1.0 | 3.6 | | 3.6 |

It is of interest to obtain similar functions by electrooptic methods which have some important advantages. In particular, electrooptic techniques have higher sensitivity, so they can be employed for studying dispersions with low number densities, at which no particle aggregates are formed. Finally, these methods permit allowance for polydispersity of a colloidal solution.

We have studied palygorskit colloidal dispersions at NaCl electrolyte concentrations varying from 0 to $7.7 \times 10^{-4}$ mol dm$^{-3}$. It was impossible to study colloidal solutions with higher concentrations because of their high electrical conductivity, which certainly is a serious disadvantage of electrooptic techniques as compared to conductometry.

Figure 4.16 gives the plots of $N(E^2)$ for a palygorskit system at various NaCl concentrations. Figure 4.17 shows $\beta(Z)$ calculated by formula (4.16); the function $f(r)$ necessary for the calculations was given in Figure 4.11a. The experimental plot of $\beta(c)$ based on the data of Figure 4.16 is presented in Figure 4.18. A comparison of these dependences yields the values of $Z$ for the systems under study. The obtained values of $Z$ and the respective NaCl concentrations are listed in Table 4.2 together with the specific surface conductivity $\kappa^\sigma = ZK$. One can easily see that though the values of $Z$ and of the medium conductivity $K$ markedly vary in the given range of NaCl concentrations, the values of $\kappa^\sigma$ remain, in fact, constant. A variation in the NaCl concentration from $10^{-4}$ to $7.7 \times 10^{-4}$ mol dm$^{-3}$, i.e., by a factor of 7.7, leads to a variation in $\kappa^\sigma$ from $4.0 \times 10^{-9}$ to $3.6 \times 10^{-9}$ $\Omega^{-1}$, or only by 10%. The values of $\kappa^\sigma$ obtained by the electrooptic method are in good agreement with the above results.[58] Some minor differences are likely to be due to the monodisperse approximation used in that work.

## 4.4. ELECTROOPTIC EFFECTS AS A FUNCTION OF APPLIED FIELD FREQUENCY

Electrooptic effects associated with anisotropy of particle polarizability depend, as was pointed out above, on the frequency $\nu$ of the applied electric field. This is so because the induced dipole moment is a function of the field frequency and because the particle orientation at low frequencies is affected by the permanent moment. Investigation of the function $N(\nu)$ gives information about the contribution of the permanent dipole moment to the particle orientation and about the frequency dependence of polarizability anisotropy.

In Figures 4.19–4.22 we present experimental mean values of $N(\nu)$ for aqueous colloidal solutions of anisaldazine, palygorskit, synthetic diamond and benzopurpurine. It is clearly seen that the dependences are quite complicated. The value of $N$ increases with frequency below $10^5$ Hz, then the functions $N(\nu)$ have a plateau up to a few hundred cycles per second. These dependences were obtained by applying reversal rectangular electric pulses. Application of sine-shaped fields produces a similar effect. For both types of field, the $N(\nu)$ profiles are similar to each other. However, rectangular pulses make the orientation patterns at different frequencies much clearer, because the variation in $N$ reflects particle re-orientation associated only with the sign reversal, while in sine-shaped fields the time variation of the field also makes a contribution.

Let us consider the mechanism of orientational ordering of particles at different frequencies of a reversible electric field. At several hundred cycles per second, the particle orientation is determined only by the induced dipole moment. No effect of the permanent moment is present, since the characteristic time of orientation relaxation is much larger than the field period.

At lower frequencies, the moment $\mu$ starts affecting the particle orientation. When the field is low ($\mu E/kT \leq 1$), the angular distribution function of the particle axes $\mathcal{W}(\theta)$ is proportional to $E^2$ for both orientations (due to the induced and to the permanent moments). The orientational ordering decreases at $\boldsymbol{\mu} \perp \gamma$ and increases at $\boldsymbol{\mu} \parallel \gamma$ ($\gamma$ is assumed to be along the axis of maximum polarizability). If the field is high enough, $\mu$ may only decrease the ordering, or the time averaged value of $\mathcal{W}(\theta)$, because, at field reversal, $\mu$ tends to re-orient the particles, and they start oscillating in a periodic field regardless of the orientations of $\mu$ and $\gamma$.

With further decrease of the field frequency, the particles still have the time to return to the initial ordering after re-orientation, which increases their ordering at $\boldsymbol{\gamma} \parallel \boldsymbol{\mu}$ and decreases it at $\boldsymbol{\gamma} \perp \boldsymbol{\mu}$. The orientation direction will become orthogonal to the initial direction at $\boldsymbol{\gamma} \perp \boldsymbol{\mu}$ and $\mu \geq \gamma E$.

Since the values of $N$ and $\mathcal{W}(\theta)$ are related through (2.8), (2.10), and (2.26), $N$ will decrease with decreasing mean value of $\mathcal{W}(\theta)$, and the change

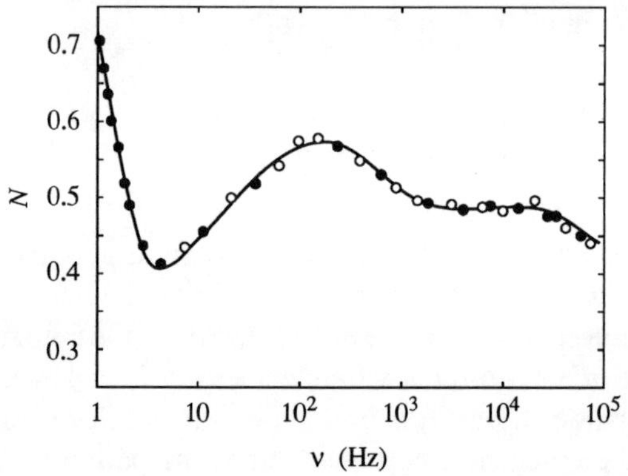


**Figure 4.19.** The plots of $N$ for an anisaldazine aqueous colloidal suspension at $E = 150$ V cm$^{-1}$. ●, in the field of rectangular pulses; ○, in the sine-shaped field.

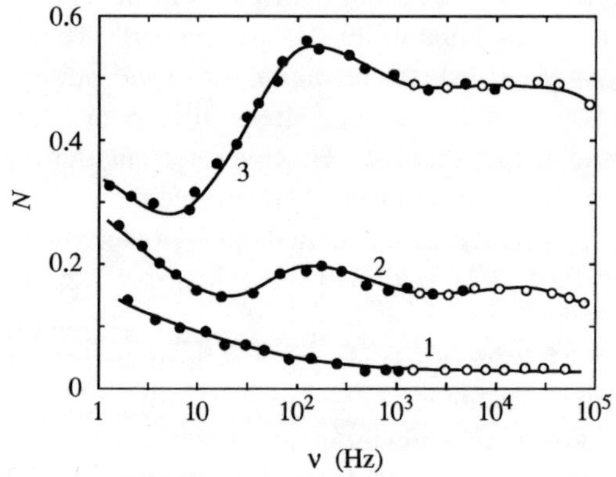


**Figure 4.20.** The $N$ curves for a palygorskit aqueous colloidal suspension at various electric field: $E = 18$ V cm$^{-1}$ (curve 1), $E = 65$ V cm$^{-1}$ (curve 2), $E = 150$ V cm$^{-1}$ (curve 3). ●, in the field of rectangular pulses; ○, in the sine-shaped field.

of the orientation direction to the orthogonal one will lead to the sign reversal of $N$. This permits an unambiguous interpretation of information inherent in the functions $N(\nu)$ and identification of the mutual orientations of $\gamma$ and $\mu$. One can see from Figures 4.19 to 4.22 that the experimental curves for $N(\nu)$ qualitatively agree with the suggested mechanism of particle orientation in the frequency range from several hundred cycles per second to zero.

In all the dispersions studied, the field $E = 150$ V cm$^{-1}$ induced a nearly complete orientation of particles. In the range of several cycles per second, the value of $N$ decreased with decreasing field frequency down to a deep minimum and then increased again. With decreasing field, the minimum depth becomes smaller and then is disappears entirely. For example, at $E = 10$ and 20 V cm$^{-1}$

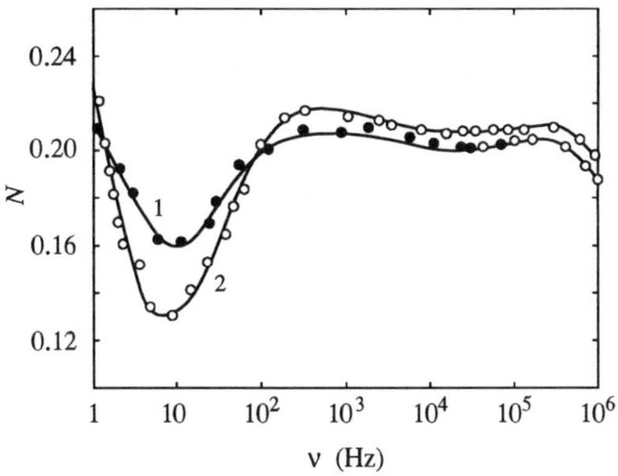


**Figure 4.21.** The $N$ curves for a diamond aqueous colloidal suspension at $E = 150$ V cm$^{-1}$) for a system with small (curve 1) and large (curve 2) particles. ●, in the field of rectangular pulses; ○, in the sine-shaped field.

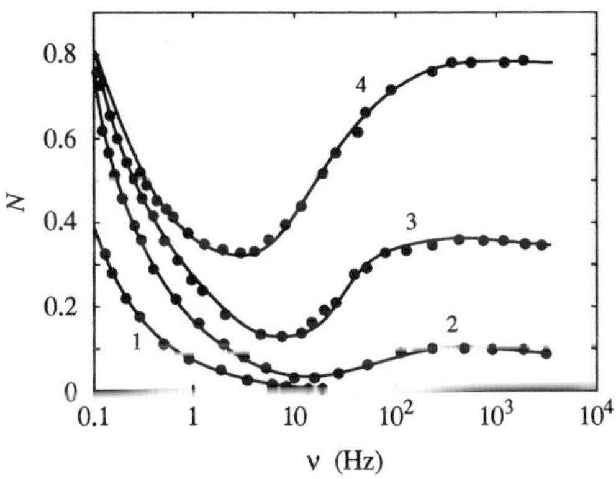


**Figure 4.22.** The $N$ curves for a benzopurpurine aqueous colloidal suspension at various electric fields. $E = 12.5$ V cm$^{-1}$ (curve 1), $E = 25$ V cm$^{-1}$ (curve 2), $E = 65$ V cm$^{-1}$ (curve 3), $E = 125$ V cm$^{-1}$ (curve 4).

the value of $N$ in benzopurpurine and palygorskit colloidal systems grows only if the external field frequency becomes lower, and at lower fields the minimum shifts towards higher frequencies. This is due to the fact that at lower $E$ only small particles have enough time for re-orientation with field reversal and their contribution to disorientation grows.

Figure 4.21 shows the $N(\nu)$ curves for diamond colloidal solutions (obtained at the same field $E = 150$ V cm$^{-1}$) with different dichroic-weighted size distribution functions. Curve 1 represents a dispersion with small particles. It is seen that the minimum $N(\nu)$ for this system is less pronounced and lies at a higher frequency than in a system with large particles. This indicates that the disorienting role of the permanent moment is greater for small particles.

Note also that in electric fields corresponding to intermediate orientation the value of $N$ is larger at very low frequencies than at high frequencies. In other words, the permanent moment makes the particle orientation more perfect and is directed in the same way as the polarizability anisotropy, namely with the long particle axis.

In the range from 1 kHz to several hundred cycles per second, one can notice a minor increase in $N$ with decreasing frequency for anisaldazine, palygorskit and diamond colloidal solutions (see Figures 4.19 to 4.21). This behavior of $N(\nu)$ in this frequency range can be explained within the theory of polarizability anisotropy.[20,66] Following this theory for the low frequency range, one should take into account the effect of electroosmotic sliding of the fluid on the particle orientation, as was pointed out in Section 3.4. In this case the expression for polarizability anisotropy can be represented as

$$\gamma_{\mathrm{tot}} = \gamma_{e0} + \gamma, \tag{4.17}$$

where $\gamma_{\mathrm{tot}}$ is the effective value of polarizability anisotropy, $\gamma_{e0}$ is the polarizability anisotropy due to the electroosmotic sliding, $\gamma$ follows from formulas (3.43) and (3.44). At fairly high frequencies, the influence of electroosmotic sliding of the fluid is negligible, since the electrolyte concentration outside the electric double layer remains practically constant when the field is applied at any particle orientation.[20] Therefore, under the conditions

$$\omega \gg \frac{2D_{\mathrm{t}}}{b^2}, \qquad \omega \gg \frac{2D_{\mathrm{t}}}{a^2}, \tag{4.18}$$

where $a$ and $b$ are the lengths of the particle semiaxes and $D_{\mathrm{t}}$ is the coefficient of translational ion diffusion, equation (4.17) takes the form

$$\gamma_{\mathrm{tot}} = \gamma. \tag{4.19}$$

When the conditions of (4.18) are not fulfilled, equation (4.19) can be used to calculate the polarizability of particles, if the $\zeta$-potential satisfies the condition[66] $\tanh(\zeta/4) \ll 1$. For the systems studied, the conditions of (4.18) are fulfilled at frequencies of the order of 1 kHz. Here one can observe a small increase in $N(\nu)$ (see Figures 4.19–4.21) caused by the increase of $\gamma$ owing to the electroosmotic sliding.

The Bulgarian research group studied the frequency dependence of the electrooptic effect $\alpha_{\mathrm{sc}}$ in sine-shaped electric fields for aqueous colloidal solutions of benzopurpurine,[59] palygorskit,[62] tobacco mosaic virus[20] and some other systems. The authors obtained correlations similar to ours for the frequency range from 15 Hz to 1 MHz. Here again, as in the case of polarizability anisotropy, it is quite unimportant which of the effects (light scattering $\alpha_{\mathrm{sc}}$ or dichroism $N$) is used in the studies. The authors believe that the reduction in the value of the electrooptic effect at low frequencies is accounted for by the contribution of the permanent dipole moment to steady-state orientation. In their opinion, the decrease of $\alpha_{sc}$ indicates that the permanent moment is orthogonal to the induced moment. The coincidence of directions of both moments should result in a larger $\alpha_{sc}$, as the field frequency decreases. This conclusion was drawn[59,62] from the theory of Jennings and Plummer[67] which describes the correlation between the light scattering anisotropy induced by a sine-shaped field and the applied field frequency. According to this theory,

$$\alpha_{sc}^{t} = \frac{\alpha_0}{G}\left[\chi + \frac{\lambda^2}{1+\tilde{\omega}^2/4} + \left(\frac{\lambda^4 + 2\lambda^2\chi}{(1+\tilde{\omega}^2/4)(1+\tilde{\omega}^2/9)} + \frac{\chi}{1+\tilde{\omega}^2/9}\right)^{-2}\right]\cos(2\tilde{\omega}\tilde{t} + \delta_0), \tag{4.20}$$

where $\tilde{\omega} = \omega/D$, $\tilde{t} = tD$ and $\lambda = \mu E/kT$ and $\chi = \gamma E^2/2kT$ were defined in Chapter 2, $\delta_0$ is the phase angle:

$$\tan\delta_0 = \frac{1}{\tilde{\omega}}\frac{\lambda^2 + (2\lambda^2 + 3\chi)(1 - \tilde{\omega}^2/4)}{5\lambda^2 + \chi(1 - \tilde{\omega}^2/4)}$$

$\alpha_0$ is the value of the electrooptic effect in a constant field;

$$G = \lambda^2 + \chi.$$

Relation (4.20) is much simplified, if the orientation is caused only by the polarizability anisotropy. Assuming $\mu = 0$, we obtain

$$\alpha_{\gamma} = \alpha_0\frac{1 + \cos(2\tilde{\omega}\tilde{t} + \delta_1)}{\sqrt{1 + \tilde{\omega}^2/9}}, \quad \tan\delta_1 = \frac{\tilde{\omega}}{3D}. \tag{4.21}$$

It is clear from equation (4.21) that $\alpha_{\gamma}$ is determined by two terms: the first term is time-independent and corresponds to the mean value over a period; the second term expresses the time dependence of $\alpha_{sc}$. At higher field frequency, the modulation amplitude drops to zero, and $\alpha_{sc}$ tends to its mean value.

If the particle orientation is determined only by the permanent moment, equation (4.20) takes the form

$$\alpha_{\mu} = \alpha_0\frac{(1 + \tilde{\omega}^2/4) + \cos(2\tilde{\omega}\tilde{t} + \delta_2)}{\sqrt{(1 + \tilde{\omega}^2/4)(1 + \tilde{\omega}^2/9)}}$$

$$\tan\delta_2 = \frac{20\tilde{\omega}}{8 - 3\tilde{\omega}^2}. \tag{4.22}$$

It also contains a time-independent term which, in contrast to (4.21), is frequency-dependent. With increasing frequency, the time averaged value of $\alpha_{\mu}$ drops and tends to zero when $\omega \to \infty$.

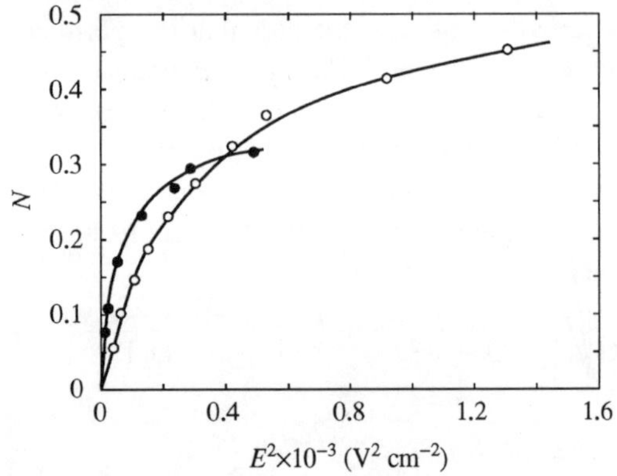


**Figure 4.23.** The $N$ curves for a synthetic diamond suspension. •, in the field of rectangular pulses at 5 kHz; ○, in the dc field.

One can see from equation (4.20) that the increase in the time-averaged $\alpha_{sc}$ at vanishing $\nu = \omega/2\pi$ testifies to parallelism between the permanent moment and the axis of maximum polarizability. The orthogonality leads to smaller $\alpha_{sc}$ values at lower frequencies. It should be stressed that equation (4.20) has been derived on the assumption that

$$\frac{\mu E}{kT} \ll 1 \quad \text{and} \quad \frac{\gamma E^2}{2kT} \ll 1. \tag{4.23}$$

The theory,[67] when applied to the analysis of $\alpha_{sc}(\nu)$ at fields which do not satisfy equations (4.23) and at too high frequencies, may become misleading as to the orientational correlation between $\gamma$ and $\mu$. The decrease of $\alpha_{sc}$ with decreasing $\nu$ in the range from 500 Hz to 15 Hz is due to the particle re-orientation at field reversal owing to their permanent moment $\mu$ rather than to the orthogonality of $\mu$ to the maximum polarizability axis. When the experiments are carried on in fields satisfying (4.23), one observes a monotonous increase of the electrooptic effect with the frequency decrease, as is shown in Figures 4.20-4.22.

Figures 4.23 and 4.24 give the plots for $N(E^2)$ in dc pulsed and alternating (5 kHz) rectangular electric fields for diamond and benzopurpurine suspensions. When dc pulses were used, their duration was minimum but sufficient for the optical effect to attain its steady-state value corresponding to the selected field strength. The described correlations show that the value of $N$ at low $E$ in the field of dc pulses is higher than in the field of 5 kHz and of the same $E$ value. This fact does not contradict the statement that $\mu \parallel \gamma$ for these dispersions.

We can infer from the experimental data presented in this Chapter that polydispersity, earlier considered as a factor which complicates the studies, can be used in electrooptic experiments for a comprehensive test of the available models of particle polarizability. The neglect of polydispersity and the use of

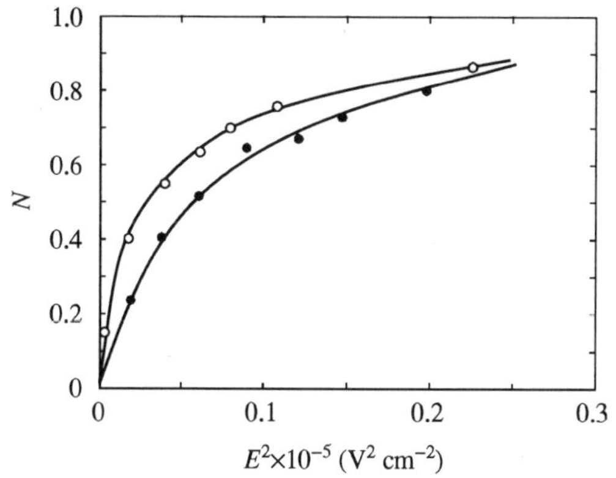


**Figure 4.24.** The $N$ curves for a benzopurpurine dispersion. ●, in the field of rectangular pulses at 5 kHz; ○, in the dc field.

the monomorphic approximation to study polydisperse systems yield distorted values of particle polarizabilities. Analysis of the monomorphic approximation shows, however, that it is justified in polarizability studies when the particle flatness or elongation are not too small.

## Chapter 5

# ELECTRICAL CONDUCTIVITY OF DISPERSIONS

## 5.1. ELECTRICAL CONDUCTIVITY MEASUREMENT IN DISPERSIONS

Investigation of the electrical double layer of particles requires knowledge of the anisotropy of polarizability $\gamma$ and its mean value $\bar{\gamma}$. The mean value can be found from the study of electrical conductivity of the dispersion. Since a dispersion is not a true solution and local conductivities in it may turn out to be different, the term "electrical conductivity of a dispersion", $\bar{K}$, should be specified.

If a dispersion with a uniform distribution of particles is placed in an experimental cell with two parallel electrode plates of an area $S$ spaced at a distance $l$, the conductivity is determined by the applied voltage $U$ and the current $i$, flowing through this inhomogeneous medium, as

$$\bar{K} = \frac{il}{SU}. \tag{5.1}$$

This value will differ from the electrical conductivity $K$ of the dispersive fluid. In the case of non-conducting particles $\bar{K} < K$, because the electric current value in expression (5.1) is determined by the electrolyte, which fills up only the interparticle spacings but not the entire space between the plates. This means that the effective area $S$ (the electric current cross section) will be smaller while the effective distance $l$ (the current flow path) will be larger, so the value of $\bar{K}$ will be lower than of $K$. At high electrolyte concentrations, the conductivity of a dispersion is proportional to that of the dispersive medium, which makes it reasonable to introduce the coefficient of structural resistivity $F$ defined by the expression:

$$F = \frac{K}{\bar{K}}. \tag{5.2}$$

For some simple cases, $F$ can be calculated if the particle sizes, shape and number density are known. At high $K$, the difference between the calculated and experimental values of $F$, according to (5.2), is not great but it becomes larger with decreasing $K$. This happens if the effect of the double layer characteristics on $F$ is neglected in the calculations. An approximate expression for $\bar{K}$ with allowance for $\kappa^{\sigma}$ and calculated $F$ has the form:[45]

$$\bar{K} = \frac{K}{F} + \frac{S_0 \kappa^{\sigma}}{(1-P)F}, \tag{5.3}$$

where $P$ is the volume fraction of particles and $S_0$ is the total surface area of the particles per unit volume of the dispersion.

Formula (5.3) should be refined, because $F$ is the same for the surface current (the second term) and for the interparticle spacings (the first term). The difference

$$\Delta K = \bar{K} - K \tag{5.4}$$

has been termed as "electrical conductivity increment". At low number densities, when the particles may be regarded as non-interacting, the value $\Delta K$ is defined by the particle number density and their dipole moments. A correlation between $\bar{K}$ and particle polarizability, though neglecting the effect of ions surrounding the particles, was derived by Maxwell. Here, however, we will present the derivation of this correlation by S. S. Dukhin,[45] because it concerns a more general case covering phenomena which arise when the ions are redistributed around a particle by an external electric field.

Let us imagine a plane dispersion layer placed in a dispersive fluid parallel to the cell electrodes. We assume the layer and the electrodes to be infinite in order to neglect the boundary effects. At the separation boundary between the two phases, the condition of current continuity must be fulfilled, because in a steady-state case the charge at the boundary remains constant:

$$E\bar{K} = E_0 K. \tag{5.5}$$

Since $\bar{K}$ differs from $K$, this equality holds only if the electric field in the dispersion, $E$, is different from that in the pure dispersive medium, $E_0$. The difference between $E$ and $E_0$ is due to the fact that under the action of the field and the flowing current the particles become polarized and release a charge onto the boundary between the dispersion and the pure dispersive medium. The relationship between the value $\Delta E = E - E_0$ and the dipole moment $d$ of a particle is defined by the well known electrostatic relationship:

$$\Delta E = -4\pi n_0 d \tag{5.6}$$

Here $n_0$ is the particle number density in the dispersion. Following equations (5.4) and (5.5), we can write

$$\Delta K = -\frac{\Delta E}{E} K. \tag{5.7}$$

If the number density is low, then:

$$\frac{\Delta E}{E} \simeq \frac{\Delta E}{E_0}. \tag{5.8}$$

Substituting into (5.7) the value of $\Delta E$ which follows from formula (5.6) and taking into account (5.8), we obtain

$$\Delta K = \frac{4\pi n_0 K d}{E_0}. \tag{5.9}$$

When $d$ is the induced moment, for randomly oriented particles

$$\bar{\gamma} E_0 = d \tag{5.10}$$

it is possible to write

$$\frac{\Delta K}{K} = 4\pi n_0 \bar{\gamma}. \tag{5.11}$$

If the particles are oriented with their maximum polarizability axis along the field, then along their orientation axis we will have

$$\Delta K_{\|} = 4\pi n_0 \gamma_{\|} K \tag{5.12}$$

and in the orthogonal direction

$$\Delta K_{\perp} = 4\pi n_0 \gamma_{\perp} K. \tag{5.13}$$

Since non-spherical particles possess anisotropy of polarizability, a system containing oriented particles will show anisotropy of the electrical conductivity increment. The difference between $\Delta K_{\|}$ and $\Delta K_{\perp}$ is equal to the difference between the conductivities in the entire system (the dispersive medium contains randomly oriented small molecules, hence, there is no anisotropy $K$):

$$\delta K = \Delta K_{\|} - \Delta K_{\perp} = \bar{K}_{\|} - \bar{K}_{\perp}.$$

Here $\bar{K}_{\|}$ and $\bar{K}_{\perp}$ are the conductivities measured along the axis of particle orientation and perpendicular to it.

Relations (5.12) and (5.13) show that the polarizability anisotropy $\gamma$ can be found not only from electrooptic studies but also from studies of electrical conductivity of dispersions with completely oriented particles. These two approaches supplement each other. For example, for very dilute dispersions it is more expedient to determine $\gamma$ from the value of the electrooptic effect, because it is easily observable at low number densities. At higher number densities in a system it is useful to apply conductometric techniques, since a concentrated system may possess multiple scattering which will complicate the observation.

Methods for measuring electrical conductivity can be classified on the basis of which parameter of the electric signal contains more information on the conductivity of the dispersion. There are two groups of methods of this kind: one for amplitude and amplitude-phase measurements and the other for field frequency measurements. In the first case the amplitude, or the amplitude and phase, of the electric current flowing through a conductometric cell are functionally related to the conductivity of the dispersion. In the second case the conductivity of the system is functionally related to the frequency of the alternating current passing through the dispersion. A frequency meter is used here as a recording device.

Conductometric methods can also be classified on an alternative basis, for instance, depending on the presence or absence of a galvanic contact of the dispersion under study with the input electric circuits of the conductivity meter. Such methods are known, respectively, as contact or noncontact techniques. Contact methods were the first ones to be used for conductivity measurements in dispersions. They are quite convenient for experimental studies and have a high accuracy. A disadvantage, however, is that the polarization of electrodes introduces an error in the measurements. Contact methods have been described comprehensively in References 68–70.

A cell used for investigation of dispersions by a contact method usually has platinum electrodes with a low polarization error (an error caused by the polarization of the electrodes). In order to reduce the polarization, which is a function of the electric current density of the electrode, it is covered with platinum black. As a result, the electrode area increases and the surface current density decreases. This is important because the latter is largely responsible for the electrode polarization. A cell with a dispersion obeys Ohm's law, and its reverse resistivity is related to the conductivity of the dispersion in the contact cell. The latter is established for each cell by using buffer electrolytic solutions of known electrical conductivity. Generally, cells are made thermostatic, since the conductivity of a dispersion is strongly dependent on its temperature.

Application of noncontact techniques allows one to avoid a polarization error and also to use aggressive media that initiate chemical reactions with the electrode material. The equipment for these methods is generally more sophisticated. However, the underlying theory of these methods is not as elaborate as for contact ones, and in some cases this circumstance makes it difficult to interpret the experimental results.

Contact methods are subdivided into direct- and alternating-current techniques according to the type of operating current.[69] Alternating-current methods can, in turn, be subdivided into high-frequency and low-frequency ones. Direct-current methods cannot be employed in the study of electrical conductivity of dispersions with a polar dispersive medium because of the distorting effect of electrophoretic motion of particles. Low-frequency methods are the ones which do not use frequencies higher than 40 or 50 kHz.

Bridge or compensating circuits, comparing circuits, and those for ordinary current measurements are used in experimental setups for investigation of electrical conductivity. The specific features of these circuits and recommendations for their application are described by B. A. Lopatin in a monograph.[69]

All low-frequency contact methods for amplitude and amplitude-phase measurements have been well elaborated and possess high precision.[70] To increase the precision of alternating-current methods, specific measures must be taken to avoid parasitic currents. For this purpose, the cell electrodes are separated from the basic volume of the electrolyte, and the electrode leads are entirely insulated from each other and from the thermostatic fluid. This provides a reduction of the experimental error in the cell constant to 0.1% and less at frequencies from 500 Hz to 20 kHz.

High-frequency amplitude conductometry commonly uses measurement of the quality of the oscillatory circuit, in which the cell serves as a capacitor, resistor or impedance. These methods are primarily employed for relative measurements in high-frequency titration. High-frequency techniques have the advantage of yielding an extremely low polarization error but they cannot give a linear dependence between the resistance of the cell electrolyte and the input signal of the measuring instrument.

Each group of methods possesses certain advantages and disadvantages. However, contact methods for sound frequencies, by which most of the available data on conductivity of dispersions have been obtained, are generally recognized as very effective, in spite of the high-precision equipment used in noncontact techniques.[69]

## 5.2. ELECTRICAL CONDUCTIVITY OF DILUTE DISPERSIONS

By dilute dispersions we understand such systems for which the volume fraction of particles $P$ satisfies the inequality $P \ll 1$ and in the expressions for $\bar{K}$ we can restrict ourselves to linear $P$ dependences of the conductivity increment. For spherical particles possessing no electric double layer, the expression for $\Delta K$ is

$$\Delta K = \frac{3}{2} PK. \tag{5.14}$$

This relation can be easily derived by substituting the expression for particle polarizability in (5.11), which in this case is a simple function of the particle radius:

$$\bar{\gamma} = -\frac{\varepsilon}{2} a^3.$$

In equation (5.14) one must also take into account that $P = 4\pi n_0 a^3/3$.

To express the effect of surface conductivity on the conductivity increment of a dispersion, it is necessary to substitute into equation (5.11) the value of polarizability, which takes into account the surface conductivity. With formula (3.27) for $\bar{\gamma}$, we can write

$$\Delta K = \frac{3}{2} PK \left(1 - \frac{3\mathrm{Rel}}{1 + \mathrm{Rel}}\right) \tag{5.15}$$

for high frequencies, at which the concentrational polarization of particles does not play an important role. Relation (5.15) makes it possible to derive for $\bar{K}$ an expression similar to (5.3). If we keep in mind that the structural resistivity of dilute dispersions with spherical particles has the form $F = 1 + 3P/2$, then

$$\bar{K} = \frac{K}{F} + \frac{9}{2} PK \frac{\mathrm{Rel}}{1 + \mathrm{Rel}}. \tag{5.16}$$

Let us compare equations (5.16) and (5.3). With the limiting transition $P \to 0$ for Rel $\ll 1$, they are identical. Otherwise we would observe a difference between these functions, because equation (5.3) does not involve the effect of surface conductivity on the total field around particles. The allowance for this fact yields (1+Rel) in the denominator of the second term in (5.16).

It is of interest to calculate the parameter $F$ for dispersions with spheroidal particles, since the model of the ellipsoid of revolution has proved fairly effective in polarizability studies. Therefore, one can suggest that spheroidal particles will be "good enough" for conductometric studies, because the conductivity increment, according to (5.11)–(5.13), varies directly with the particle polarizability.

For dilute suspensions, when one cannot neglect the nonlinear effect of $P$ on $F$, the expression for $F$ has the form:[45]

$$F = 1 + \xi \frac{P}{1+P}.$$

The quantity $\xi$ is defined by the depolarization factors $A_x$, $A_y$, and $A_z$, which are given in (3.19)–(3.21) for prolate spheroids. For oblate spheroids, instead of equation (3.21), we should write

$$A_x = \frac{p^2}{p^2-1} - \frac{p^2}{(p^2-1)^{3/2}} \left( \frac{\pi}{2} - \arctan \frac{1}{(p^2-1)^{1/2}} \right).$$

The relation between $A_x$, $A_y$, $A_z$, and $\xi$ is described by the expression for randomly orientated ellipsoids

$$\xi = \frac{1}{3} \left( \frac{1}{1-A_x} + \frac{1}{1-A_y} + \frac{1}{1-A_z} \right).$$

However, for low or high values of $p$ the expression for $\xi$ takes a simple form:[45] $\xi = 1 + 0.21p$ at $p \ll 1$ and $\xi = 5/3$ at $p \gg 1$. The mean values of polarizability in this case can be found from the formula:[20]

$$\bar{\gamma} = \frac{1}{3}\gamma_{\|} + \frac{2}{3}\gamma_{\perp}. \tag{5.17}$$

To obtain the dependence of the conductivity increment on the surface conductivity of particles it is sufficient to combine equations (5.17), (5.11), (3.43), and (3.44). For prolate spheroids we can write

$$\Delta K = \frac{PK}{3} \left\{ \left( \frac{3}{2} I_{\|} \text{Rel} - 1 \right) \left[ 1 + A_x \left( \frac{3}{2} I_{\|} \text{Rel} - 1 \right) \right]^{-1} \right.$$
$$\left. + 2 \left( \frac{3}{2} I_{\perp} \text{Rel} - 1 \right) \left[ 1 + \frac{1-A_x}{2} \left( \frac{3}{2} I_{\perp} \text{Rel} - 1 \right) \right]^{-1} \right\}. \tag{5.18}$$

In the case of a polydisperse system, where the particle size distribution is described by the function $\varphi(r)$, formula (5.18) can be generalized

$$\Delta K = \frac{4\pi}{9} K \int_0^\infty r^3 \Psi \kappa^\sigma \varphi(r) \mathrm{d}r,$$

where

$$\Psi = \left(\frac{3}{2} I_\| \frac{Zp^{1/3}}{r} - 1\right)\left[1 + A_x \left(\frac{3}{2} I_\| \frac{Zp^{1/3}}{r} - 1\right)\right]^{-1} + 2\left(\frac{3}{2} I_\perp \frac{Zp^{1/3}}{r} - 1\right)\left[1 + \frac{1 - A_x}{2}\left(\frac{3}{2} I_\perp \frac{Zp^{1/3}}{r} - 1\right)\right]^{-1}.$$

For oblate spheroids or rod-shaped particles, the expressions for $\Delta K$ can be derived in a similar way by substituting the expressions for $\gamma_\|$ and $\gamma_\perp$ for particles of these shapes into equation (5.17). For example, for needle-shaped particles in a monodisperse system ($p \gg 1$), one can get

$$\bar{K} = \frac{K}{F} + \frac{S_0 \kappa^\sigma}{3}\left(\frac{1}{1 + Z/rp^{2/3}} + \frac{2}{1 + 2Z/rp^{-1/3}}\right). \tag{5.19}$$

For oblate spheroids ($p \ll 1$), we have

$$\bar{K} = \frac{K}{F} + \frac{2\kappa^\sigma S_0}{3}\left(\frac{1}{1 + Z/rp^{1/3}} + \frac{4}{1 + Z/rp^{-2/3}}\right). \tag{5.20}$$

Relations (5.19) and (5.20), in contrast to (5.3), do not contain the parameter $F$ in the second term.

At high electrolyte concentrations the effect of $\kappa^\sigma$ is negligible and we can assume $\bar{K} = K/F$, $\Delta K < 0$, whereas at lower concentrations $\Delta K$ increases and crosses the abscissa at $\Delta K = 0$. The case when the decrease in the conductivity due to $F$ is compensated by its increase due to $\kappa^\sigma$ corresponds to the isoconductivity of a dispersion. In a polydisperse system the isoconductivity occurs at different concentrations of the electrolyte for different groups of particles. This is similar to electrooptic studies, when isopolarity does not occur simultaneously throughout the system.

The values of $\kappa^\sigma$ for clays derived from (5.3) satisfactorily agree with electrooptic data and lie, as a rule, in the limits $(0.6–6)\times 10^{-9}\,\Omega^{-1}$ for the conductivity variation within $10^{-6}–10^{-3}\,\Omega^{-1}\,\mathrm{cm}^{-1}$. However, in some cases $\kappa^\sigma$ changes non-monotonously with increasing electrolyte concentration, which seems to be associated with particle aggregation. In order to find evidence for the particle interaction, it is necessary to follow the evolution of the relaxation times of electrooptic effects or of the conductivity anisotropy induced by an external electric field (see Chapter 4). S. S. Dukhin[45] presents a systematic discussion of data on the effect of electrolyte concentrations on the conductivity of suspensions.

The conductivity increment is determined by the dipole moments of particles, so there must exist a frequency dependence of conductivity when the frequency changes the moments. For dilute dispersions, these dependences are easy to describe because the conductivity increment is proportional to the particle polarizability. One of the dependence regions corresponds to the frequencies:

$$\omega_1 \ll \frac{2D_e}{a^2}.$$

In this region the conductivity increment for spheres can be described by formula (3.30). Then the expression for $\Delta K$ has the form:

$$\Delta K = \frac{3}{2} PK \left(1 - \frac{6\mathrm{Rel}}{1 + 4\mathrm{Rel}}\right).$$

At higher frequencies the concentrational polarization gradually diminishes its effect on the conductivity and its increment is described by formula (5.15). The other dependence region corresponds to the frequencies:

$$\omega_2 \ll 2D_e\kappa^2, \tag{5.21}$$

at which we can observe only a Maxwell–Wagner polarization. To describe $\Delta K$ in this region, one should make use of formulas (3.37) and (3.38). The range of frequencies

$$\omega_1 \ll \omega \ll \omega_2 \tag{5.22}$$

may be quite large for some dispersions. For example, in aqueous systems with particle sizes of about 1 $\mu$m the concentrational polarization occurs at frequencies less than 1 kHz, while the relaxation of an electric double layer manifests itself at frequencies of several hundred kilocycles per second.

## 5.3. ELECTRICAL CONDUCTIVITY OF CONCENTRATED DISPERSIONS

The theory of electrical conductivity of a dilute dispersion relates the double layer parameters to the electrical conductivity increment. However, in some dilute systems the appropriate measurements are difficult to make. Sometimes the particles settle down to the cell bottom and interact with each other. The conductivity increment cannot often be measured reliably, because here we deal with small differences between large numbers, one of which is the electrical conductivity of the dispersion and the other is the conductivity of the dispersive medium. This circumstance makes the researcher use a colloidal system with a high number density. Concentrated clay dispersions have been analyzed by S. Dukhin.[45]

Numerical calculations of the double layer parameters became feasible in the 1980s after a conductivity theory for concentrated dispersions had been suggested. This theory made allowance for the nonlinear dependence of the conductivity increment on the volume fraction of particles. It should be noted that this theory is still far from being completed since it deals only with spheres and cylinders in a thin double layer approximation.

The conductivity theory for concentrated dispersions is based on a cellular model,[71,72] in which one considers a small volume of a dispersion with one particle in it, or a dispersion "cell", instead of the whole of the system. This permits analysis of the basic transfer phenomena occurring in concentrated dispersions by solving a much simplified problem. In this case the interaction effects with the neighboring particles are taken into account by giving the appropriate boundary conditions for the cell. The cell is supposed to have the same shape as the particle, and the cell size is chosen on the assumption that the total volume fraction of particles in the system is equal to the volume ratio of the particle and the cell. For example, if the volume fraction of spherical particles of radius $a$ in a dispersion is $P$, then the cell is a sphere of radius $b = aP^{-1/3}$. The cell sizes for cylindrical particles are defined in a similar way. It is clear that the cellular model is well applicable for description of processes in real dispersions only if the number density is the same throughout the entire volume of the system. Otherwise one would have to choose a large number of cells, each with its own boundary conditions, which would make the problem practically unsolvable.

The electrical conductivity of a cell $\tilde{K}$ is defined as $\tilde{K} = \bar{j}/\bar{E}$ where $\bar{j}$ is the mean current density throughout the cell, $\bar{E}$ is the total field in the cell. The full expression for $\tilde{K}$ was given by Yu. Borkovskaya.[73] Here we will confine ourselves to simplified formulas which do not incorporate the osmotic components of the surface current in capillaries. They do not strongly affect the number density behavior of the cell conductivity but complicate the mathematical calculations.

At high electrolyte concentrations, when the surface conductivity has no significant effect on $\bar{K}$, the cellular model gives the same result as the Maxwell–Wagner theory. For a suspension of spherical particles

$$\Delta K = \frac{3}{2}K\frac{P}{1+P/2}.$$

The influence of concentrational polarization of the double layer on the conductivity of a suspension of dielectric spheres is described by the formula[73]

$$\Delta K = \frac{3}{2}K\frac{PR_1}{1+PR_1/2}, \tag{5.23}$$

where

$$R_1 = \frac{1+P/2-(1+2P)\mathrm{Rel}}{1+P/2+(2-P/2)\mathrm{Rel}}, \tag{5.24}$$

while for cylinders we can write

$$\Delta K = -2K \frac{PR_2}{1 + PR_2}, \tag{5.25}$$

where

$$R_2 = \frac{1 + P + 2P\text{Rel}}{1 + P + 2\text{Rel}}. \tag{5.26}$$

Let us consider the basic processes responsible for this behavior of the function $\Delta K(P)$. The contribution of particles to the conductivity is determined by the number density and the induced dipole moment $d$ of each particle. The induced moment is formed by the internal field of the dispersion and is proportional to the applied field polarizing the particles and inducing an electric current in the system. When the volume fraction of particles is small, this electric field is practically the same as in the cell. A similar assumption was employed in Section 5.1 to derive the relation in (5.9). With increasing volume fraction of particles, the polarizing field will differ to a greater extent from the average cell field due to the Lorentz field contribution following from the relationship

$$E_\text{L} = \frac{4\pi}{3} n_0 d.$$

If Rel is close to zero, the dipole moment will be opposite in sign to the external field. The electrical conductivity of the dispersion in this case will decrease more slowly than in the linear law for an increasing volume fraction. If Rel is much larger than unity, the Lorentz field does not attenuate but, on the contrary, amplifies the external field, because the dipole moments in this case are aligned with the field. As a result, the conductivity increment increases more rapidly than it does in the linear law, as the volume fraction of particles increases.

Concentrational polarization is of particular importance in measuring electrical conductivity at lower frequencies, because it reduces the positive contribution of surface conductivity to the total conductivity of a dispersion. Dipole moments induced by concentrational polarization are opposite in sign to the external field. With increasing volume fraction of particles, the concentrational fields of the neighboring particles begin to overlap, and the region of higher electrolyte concentration around the chosen particle is superimposed with the regions of lower concentration of the neighboring particles and vice versa. This reduces the effect of concentrational polarization on the $\bar{K}$ value as the volume fraction of particles increases. In some cases, as $P$ rises, such a behavior of the polarization dependence on the volume fraction leads to a situation when the conductivity $\bar{K}$ reaches a minimum and starts to increase in spite of the fact that Rel remains practically unchanged. Another specific feature of $\tilde{K}$ caused by the nonlinear $P$ dependence of $K$ is that Rel does depend on the volume fraction of particles at the isoconductivity point. This is not the case with dilute suspensions. For example, it follows from formulas (5.23)–(5.26)

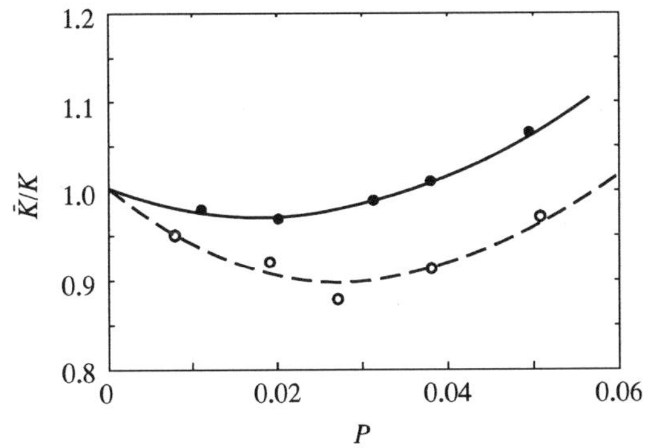


**Figure 5.1.** A plot of relative electrical conductivities $\bar{K}/K$ in a palygorskit suspension versus the volume fraction $P$ of $NaNO_3$ aqueous solution at $NaNO_3$ concentration $5 \times 10^{-4}$ mol dm$^{-3}$ (solid curve) and with polyvenylsterol addition of 0.3 mg per 1 m$^2$ adsorption surface (dashed curve).

that for spherical particles at the isoconductivity point

$$\mathrm{Rel}_{\mathrm{iso}} = \frac{(1 + P/2)}{(1 + 2P)}$$

and for cylinders orthogonal to the field

$$\mathrm{Rel}_{\mathrm{iso}} = \frac{1}{2}\left(1 + \frac{1}{P}\right).$$

The conclusions drawn from the conductivity theory for concentrated systems were tested in studies of the conductivity increment as a function of the volume fraction of particles in a palygorskit aqueous suspension. The results are shown in Figure 5.1. One can see a qualitative confirmation of the conclusion that $\Delta K$ passes through a minimum.[73]

For frequencies $\omega \gg \omega_1$, at which the effect of concentrational polarization becomes negligible, the application of the cellular model can be substantiated as follows. There is no polarizational variation of the concentration, and the only factor determining the conductivity increment of the system is the local value of the specific surface areas and their distribution. In this case the frequency dependence arises only with the relaxation of a free polarization charge at a frequency defined by (5.21). As for spheres, the conductivity increment in the region described by the inequalities of (5.22) is

$$\Delta K = K \frac{1 - 3P(1 - 2\mathrm{Rel})}{1 + 2\mathrm{Rel} + P(1 - 2\mathrm{Rel})}.$$

At these frequencies $\mathrm{Rel}_{\mathrm{iso}}$ no longer depends on $P$, then for spheres $\mathrm{Rel}_{\mathrm{iso}} = 1/2$ and for cylinders $\mathrm{Rel}_{\mathrm{iso}} = 1$.

# Chapter 6

# ELECTRICAL CONDUCTIVITY ANISOTROPY OF POLYDISPERSE SYSTEMS

## 6.1. INTRODUCTION

The conductivity measurement in polydisperse systems can provide only tentative information on the electrical characteristics of particles, because it yields a size-averaged value of $\bar{\gamma}$. In order to avoid a possible distortion of double layer measurements due to the system polydispersity, it is necessary to measure the conductivity anisotropy in variable orienting fields.[74] To our knowledge, this approach was not used before 1985, except perhaps by G. Schwarz who studied the conductivity variation in a polymer solution.[75] Systematic studies of this problem seem to have been hampered by a lack of conductometric theory for high fields and by the difficulties of doing an experiment in which the orienting field affects the measuring devices and distorts the data. One can circumvent this problem by using electrooptic methods. However, their application to highly concentrated systems has a disadvantage, because these methods do not detect the secondary light scattering by particles. Obviously, a combination of orientational and conductometric techniques can make the study of dispersions with small interparticle spacings quite effective.

Lately, there has been much interest in the research on particle interactions and coagulation. This is evidenced by a fairly large number of publications on the stability of various dispersions.[2,9,76,77] Studies on the formation of aggregates and electrically induced aggregation can be performed by measuring the anisotropy of the system conductivity when the particles are orientationally ordered.

Another advantage of orientational and conductometric methods over electrooptic techniques is the possibility to build a rigorous theory for arbitrary particle sizes, while in electrooptic studies one has to resort to approximate theories of light scattering. Analysis of the conductivity dependence on the orienting field strength may provide the same information about the double layer characteristics in polydisperse systems as electrooptic studies.

Since the underlying theory of the combined orientational and conductometric approach has not been fully described in the literature, we will discuss it in some detail in this book.

## 6.2. THEORY OF FIELD-INDUCED ANISOTROPY OF ELECTRICAL CONDUCTIVITY IN DISPERSIONS

Let us consider the relationship between the anisotropy of electrical conductivity in a dispersion and the degree of orientational ordering of particles induced by an applied electric field.

Suppose that the particles are oriented by a high frequency sine-shaped field, so that we can ignore the effect of the permanent dipole moment on the orientation and on the induced macroscopic anisotropy in a steady-state case. The particles are oriented by the field $\boldsymbol{E}$ (orienting field), while the electrical conductivity of the dispersion is measured by another field $\boldsymbol{E}_{\text{meas}}$ (measuring field). We assume the latter to be so weak that it does not actually contribute to the particle orientation. Let us find a correlation between the relative variation in the system conductivity in various directions and the orienting field strength. The quantity $\gamma$ defines, as previously, the difference in the particle polarizabilities.

In this analysis we assume all the particles to be identical spheroids made up of the same material. In the coordinate system with the basic vectors $\boldsymbol{e}_1$, $\boldsymbol{e}_2$ and $\boldsymbol{e}_3$ along the principal axes of an ellipsoid (eigensystem of coordinates), the polarizability tensor is diagonal and its principal values are equal to $\gamma_1$, $\gamma_2$ and $\gamma_3$. Let the value of maximum $\gamma_1$ correspond to the direction of maximum polarizability of a particle. Now we choose a laboratory Cartesian system of coordinates $x$, $y$, $z$ with the basic vectors $\boldsymbol{e}_x$, $\boldsymbol{e}_y$, $\boldsymbol{e}_z$. When the coordinates of this particle are turned through the corresponding Euler angles, they will coincide with the Cartesian coordinates. The measuring cell and the dispersion itself are designed in such a way that neither the external field nor the properties of the dispersion vary within the region of the conductivity measurement.

Let us introduce the rotation angles $\theta$, $\alpha$, and $\varphi$ of the eigenbasis of a particle with respect to the laboratory system of coordinates with $\theta$ as an angle between the vectors $\boldsymbol{e}_1$ and $\boldsymbol{e}_x$. The angle $\theta$ varies in the range $0 \le \theta \le \pi$; the other two Euler angles, $\alpha$ and $\varphi$, vary from 0 to $2\pi$. The expected effect will be described in the laboratory system of coordinates. To this end, the eigenbasis of the particle $\boldsymbol{e}_1$, $\boldsymbol{e}_2$, and $\boldsymbol{e}_3$ should be expressed in terms of the basis $\boldsymbol{e}_x$, $\boldsymbol{e}_y$ and $\boldsymbol{e}_z$, respectively. The matrix of transformation to the eigenbasis of the particle is denoted as $\hat{C}$. It characterizes the orthogonal transformation of three-dimensional Euclidean space into itself. In this case $\hat{C}^{-1} = C^*$ and the matrix $\hat{C}$ can be represented as

$$\hat{C} = \begin{pmatrix} -\sin\theta\sin\varphi & \begin{matrix}\cos\alpha\cos\theta\sin\varphi \\ +\cos\alpha\cos\varphi\end{matrix} & \begin{matrix}-\sin\alpha\cos\theta\sin\varphi \\ +\cos\alpha\cos\varphi\end{matrix} \\ -\sin\theta\cos\varphi & \begin{matrix}\cos\alpha\cos\theta\cos\varphi \\ -\sin\alpha\sin\varphi\end{matrix} & \begin{matrix}-\sin\alpha\cos\theta\cos\varphi \\ -\cos\alpha\sin\varphi\end{matrix} \\ \cos\theta & \sin\theta\cos\alpha & -\sin\theta\sin\alpha \end{pmatrix} \tag{6.1}$$

The basis of the laboratory system of coordinates yields the eigenbasis of the particle from the transformation $\boldsymbol{e} = \boldsymbol{e}_{x,y,z}\hat{C}$, where $\boldsymbol{e} = \boldsymbol{e}_1 + \boldsymbol{e}_2 + \boldsymbol{e}_3$ and $\boldsymbol{e}_{x,y,z} = \boldsymbol{e}_x + \boldsymbol{e}_y + \boldsymbol{e}_z$.

If the measuring field is along the axis $z$, the vector

$$\boldsymbol{E}_{\text{meas}} = E_{\text{meas}}\,\boldsymbol{e}_z$$

in the eigenbasis of the particle takes the form

$$\boldsymbol{E}'_{\text{meas}} = \hat{C}^{-1}\boldsymbol{E}_{\text{meas}} = \hat{C}^{*}\boldsymbol{E}_{\text{meas}}. \tag{6.2}$$

By substituting equation (6.1) into equation (6.2), we get

$$\boldsymbol{E}'_{\text{meas}} = \begin{pmatrix} \cos\theta \\ \sin\theta\cos\alpha \\ -\sin\theta\sin\alpha \end{pmatrix} \boldsymbol{E}_{\text{meas}}. \tag{6.3}$$

In the eigenbasis of the particle, its polarization vector $\boldsymbol{P}'$ equals

$$\boldsymbol{P}' = \hat{\gamma}\boldsymbol{E}'_{\text{meas}}, \tag{6.4}$$

where $\hat{\gamma}$ is the matrix of the particle polarizability

$$\hat{\gamma} = \begin{pmatrix} \gamma_1 & 0 & 0 \\ 0 & \gamma_2 & 0 \\ 0 & 0 & \gamma_3 \end{pmatrix}. \tag{6.5}$$

Now we substitute (6.5) into (6.4) and average it over the angles $\alpha$ and $\varphi$, assuming the distribution of the particle axes over these angles to be isotropic. The averaged value of the induced dipole moment of the particle for a given measuring field is denoted as $\langle \boldsymbol{P} \rangle_{\alpha,\varphi}$:

$$\langle \boldsymbol{P} \rangle_{\alpha,\varphi} = \begin{pmatrix} 0 \\ 0 \\ \left(\gamma_1 - \frac{\gamma_2+\gamma_3}{2}\right)\cos^2\theta + \frac{\gamma_2+\gamma_3}{2} \end{pmatrix} \boldsymbol{E}_{\text{meas}}.$$

To find the mean polarizability of the particles oriented along the $z$-axis, it is sufficient to divide $\langle P \rangle_{\alpha,\varphi}$ by $E_{\text{meas}}$. The result is

$$\langle \gamma \rangle^{z}_{\alpha,\varphi} = \left(\gamma_1 - \frac{\gamma_2+\gamma_3}{2}\right)\cos^2\theta + \frac{\gamma_2+\gamma_3}{2}. \tag{6.6}$$

For an entirely random distribution of particle axes at the mean value of $\cos^2\theta$, the expression

$$\cos^2\theta = \frac{1}{3}$$

is valid, and the mean polarizability along the $z$-axis is described by

$$\bar{\gamma} = \frac{\gamma_1+\gamma_2+\gamma_3}{3}.$$

The polarizability along the $x$- and $y$-axes will have the same value, which follows from the assumption of spherical symmetry of the axis distribution for random particle orientation. If the measuring field is directed along the $x$- or $y$-axis, the polarizabilities averaged over the angles $\alpha$ and $\varphi$ are described in a way similar to (6.6):

$$\langle\gamma\rangle^x_{\alpha,\varphi} = \langle\gamma\rangle^y_{\alpha,\varphi} = \frac{1}{2}\left(\gamma_1 - \frac{\gamma_2+\gamma_3}{2}\sin^2\theta + \frac{\gamma_2+\gamma_3}{2}\right). \tag{6.7}$$

Let us denote $\gamma_{\|} = \gamma_1$, $\gamma_{\perp} = (\gamma_2+\gamma_3)/2$, then

$$\bar{\gamma} = \frac{1}{3}\gamma_{\|} + \frac{2}{3}\gamma_{\perp}.$$

For entirely disoriented particles we can write[20]

$$\Delta K = \frac{4\pi n_0}{\epsilon}\left(\frac{1}{3}\gamma_{\|} + \frac{2}{3}\gamma_{\perp}\right)K. \tag{6.8}$$

Now we will analyze the variation in the conductivity increment with particle orientation. The vector of the orienting field $\boldsymbol{E}$ is assumed to be along the $z$-axis. The conductivity variation due to the particle orientation along the $z$-axis is denoted as $\delta K_{\|}$, and in the direction orthogonal to this axis as $\delta K_{\perp}$, i.e. $\delta K_{\|} = K_{\|} - \bar{K}$, $\delta K_{\perp} = K_{\perp} - \bar{K}$, where $K_{\|}$ and $K_{\perp}$ are the conductivities along the orienting field $\boldsymbol{E}$ and orthogonal to it. Obviously, the conductivity variation of the dispersion is related, as was noted above, to the variation of its increment:

$$\delta K_{\|} = \Delta K_{\|} - \Delta K, \qquad \delta K_{\perp} = \Delta K_{\perp} - \Delta K.$$

The expressions for $\Delta K_{\|}$ and $\Delta K_{\perp}$ can be obtained by substituting, in the increment formula, the expressions for the polarizabilities along and perpendicular to the $z$-axis, respectively, which were averaged over the angle $\theta$ formed by the maximum polarizability direction and the orienting field vector. Since the field $\boldsymbol{E}$ is non-zero, the distribution of the particle axes over the angle $\theta$ is not isotropic. However, for the angular distribution function of the particle axes one can use formula (1.9) in a way similar to derivation of $N$ (see Chapter 2). The influence of the permanent dipole moment of particles on their orientation is neglected, since the electric field has a high frequency. Thus, for the distribution function of the particle axes over the angle $\theta$, we can write

$$F(\theta) = \exp\left(\frac{\gamma\cos^2\theta E^2}{2kT}\right)\left[\int_0^{\pi}\exp\left(\frac{\gamma\cos^2\theta E^2}{2kT}\right)\sin\theta d\theta\right]^{-1}. \tag{6.9}$$

In this case

$$\delta K_{\|} = \frac{4\pi n_0}{\epsilon}K\int_0^{\pi}[\langle\gamma\rangle^z_{\alpha,\varphi}F(\theta) - \bar{\gamma}]\sin\theta d\theta. \tag{6.10}$$

After the substitution of formulas (6.6) and (6.9) into (6.10) and integration, we have

$$\delta K_{\|} = \frac{4\pi n_0}{\epsilon}(\gamma_{\|} - \gamma_{\perp}) \left( \frac{\int_0^1 x^2 \exp(\chi, x^2)\mathrm{d}x}{\int_0^1 \exp(\chi, x^2)\mathrm{d}x} - \frac{1}{3} \right) K, \tag{6.11}$$

where $\chi$ is defined, similarly to equation (2.11), by the expression:

$$\chi = \frac{\gamma_{\|} - \gamma_{\perp}}{2kT} E^2.$$

Equation (6.11) coincides, with the accuracy to a constant factor, with (2.9) for steady-state electrooptic effects. Indeed, with formula (2.10) and taking $\lambda = 0$, we obtain

$$\delta K_{\|} = \frac{8\pi n_0}{3\epsilon} K\gamma\mathcal{F}(\chi, 0). \tag{6.12}$$

The value $\delta K_{\perp}$ can be found by the formula

$$\delta K_{\perp} = \frac{4\pi}{\epsilon} n_0 K \int_0^{\pi} \left[ \langle\gamma\rangle^{z}_{\alpha,\varphi} F(\theta) - \bar{\gamma} \right] \sin\theta \mathrm{d}\theta.$$

Like for $\delta K_{\|}$, we substitute (6.7) and (6.9) into this expression to get:

$$\delta K_{\perp} = -\frac{4\pi}{3\epsilon} n_0 K\gamma\mathcal{F}(\chi, 0). \tag{6.13}$$

When comparing formulas (6.12) and (6.13), it is easy to see a simple relationship between $\delta K_{\|}$ and $\delta K_{\perp}$

$$\delta K_{\|} = -\frac{1}{2}\delta K_{\perp}. \tag{6.14}$$

A comparison of formulas (6.12), (6.13), (1.13), and (2.9) shows that the $E$ dependences of $\delta K_{\|}$, $\delta K_{\perp}$ and $\Delta A/A$ coincide with an accuracy to a constant factor. However, expressions (6.12) and (6.13) are valid for particles of arbitrary sizes, whereas equations (1.13) and (2.9) are applicable only to particles of definite sizes. Relations (1.13) and (2.9) describe dispersions with a much narrower range of number densities than (6.12) and (6.13). For the last two expressions it is sufficient that the conductivity increment varied directly with the number density. Relation (6.14) can be used even for very concentrated dispersions, provided that the size distribution function $F(\theta)$ possesses an axial symmetry.

For polydisperse systems that can be described by the size distribution function $\varphi(r)$, the conductivity variation $\delta K_{\|,\perp}$ can be represented, similarly to an electrooptic effect, by the formula

$$\delta K_{\|,\perp} = \int_0^{\infty} \delta K_{\|,\perp}(r)\varphi(r)\mathrm{d}r. \tag{6.15}$$

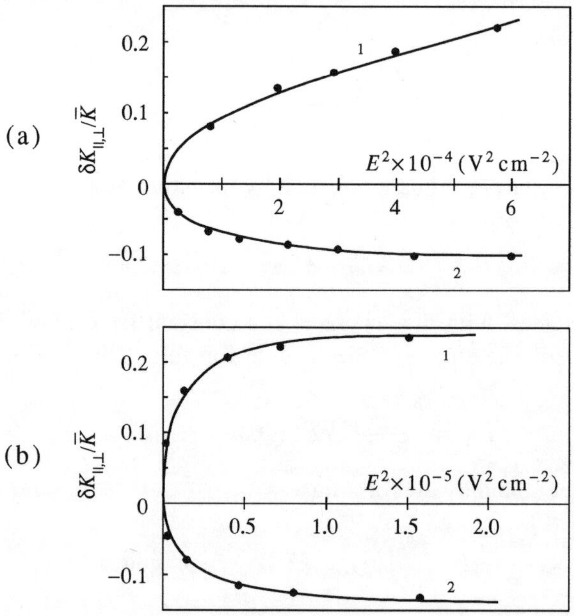


**Figure 6.1.** The plots of $\delta K_{\parallel}/\bar{K}$ (curves 1) and $\delta K_{\perp}/\bar{K}$ (curves 2) versus $E^2$ for aqueous colloidal dispersions. (a) Solution of carboxyl iron. (b) Solution of aniline brown.

It is clear from equations (6.14) and (6.15) that the correlation between $\delta K_{\parallel}$ and $\delta K_{\perp}$ for a polydisperse system is the same as for a monodisperse one.

To estimate quantitatively $\delta K_{\parallel,\perp}$, we will consider the polarizability model for spheroids described by equations (3.43) and (3.44). When the semiaxis ratio of spheroids varies from 1 to 3, the maximum variation in the particle conductivity corresponding to their complete orientation is related to the volume fraction $P$ as follows

$$0 < \delta K_{\parallel} < 4.5PK \tag{6.16}$$

$$-2.3PK < \delta K_{\perp} < 0. \tag{6.17}$$

In case of particle aggregation, all the above expressions, except for (6.14) and (6.15), may become quite different.

It should be noted that for small differences between the conductivity of a dispersion $\bar{K}$ and the conductivity of a dispersive medium $K$, we can write:

$$\frac{\delta K_{\parallel,\perp}}{\bar{K}} = \frac{\delta K_{\parallel,\perp}}{K} - O\left(\frac{\delta K_{\parallel,\perp}\Delta K_{\parallel,\perp}}{K^2}\right).$$

Using this approximation, we have, instead of equations (6.12) and (6.13), the expressions

$$\frac{\delta K_{\parallel}}{\bar{K}} = \frac{8\pi}{3\epsilon} n_0 \gamma \mathcal{F}(\chi, 0), \qquad \frac{\delta K_{\perp}}{\bar{K}} = -\frac{4\pi}{3\epsilon} n_0 \gamma \mathcal{F}(\chi, 0). \tag{6.18}$$

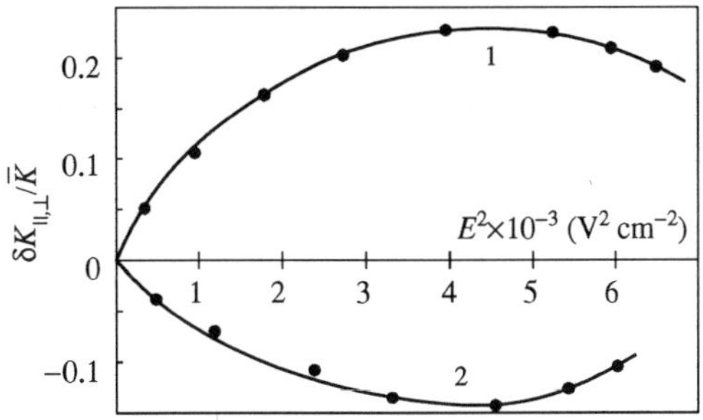


**Figure 6.2.** The plots of $\delta K_{\parallel}/\bar{K}$ (curve 1) and $\delta K_{\perp}/\bar{K}$ (curve 2) versus $E^2$ at $p = 0.025$ for an aqueous suspension of carbon black.

The term $O\left(\frac{\delta K_{\parallel,\perp}\Delta K_{\parallel,\perp}}{K^2}\right)$ has been neglected here.

Experimental determination of $\delta K_{\parallel}/\bar{K}$ and $\delta K_{\perp}/\bar{K}$ involves relative changes in the electrical conductivity which can be measured much more accurately than the absolute conductivities. Figure 6.1 illustrates the relative variations in the conductivity of aqueous colloidal solutions of aniline brown and carboxyl iron as a function of the squared orienting field. Experimental methods for the study of conductivity anisotropy is discussed in Section 6.3. The volume fraction of particles in the colloidal solutions was 5%. One can see from Figure 6.1 that the relative variation in the conductivities is in agreement with the estimates from (6.16) and (6.17), and the experimental values of $\delta K_{\parallel}/\bar{K}$ and $\delta K_{\perp}/\bar{K}$ confirm the theoretical expression (6.14) in the entire variation range.

Figure 6.2 illustrates the validity of (6.14) beyond the applicability of equation (6.18) with reference to $\delta K_{\parallel,\perp}/\bar{K}$ as a function of $E^2$ for an aqueous suspension of finely dispersed carbon soot with a volume fraction of 2.5%. The curves are non-monotonic because of the formation of particle aggregates in the electric field. This disagrees with the formulas of (6.18), as the experimental anisotropy of electrical conductivity noticeably exceeds its estimates from equations (6.16) and (6.17). However, Figure 6.2 shows that equation (6.14) still holds valid, which indicates a large number of aggregates in the dispersion. Relation (6.14) will be violated if the dispersion is restructured to form a homogeneous medium. Thus, formula (6.14) can serve as a qualitative indicator of aggregate density in a dispersion.

## 6.3. EFFECT OF PARTICLE AGGREGATION ON ELECTRICAL CONDUCTIVITY ANISOTROPY

A simple indication of particle aggregation in conductometric studies is a discrepancy between the values of $\delta K_{\parallel,\perp}/\bar{K}$ calculated for non-interacting particles and experimental values. This is equally true for the conductivity increment $\Delta K$. Variation in $\Delta K$ and $\delta K_{\parallel,\perp}$ is due to the difference between the polarizability of an aggregate and the sum of individual polarizabilities of particles that comprise the aggregate. However, if the charge of interacting particles does not vary noticeably, their surface conductivity does not change much either, which means that the increment $\Delta K$ cannot show a remarkable variation during particle aggregation. Therefore, the above method will become inapplicable.

A more accurate and sensitive way of detecting particle aggregation is offered by the analysis of the relaxation times of conductivity anisotropy. The anisotropy relaxation is described by the expression:

$$\frac{\delta K_{\parallel,\perp}(t)}{\bar{K}} = \left(\frac{\delta K_{\parallel,\perp}}{\bar{K}}\right)_0 \exp(-6Dt),$$

where $(\delta K_{\parallel,\perp}/\bar{K})_0$ is the relative variation in electrical conductivity for steady-state orientation. The conductivity is described by formula (6.18). However, the formation of pair aggregates will increase the relaxation time by a factor of 6–8. Such a great change in the time variation of $\delta K_{\parallel,\perp}/\bar{K}$ during the relaxation makes it very perspective for particle aggregation studies. For instance, in the plots for $\delta K_{\parallel,\perp}/\bar{K}$ given in Figure 6.2 the relaxation time of induced anisotropy in low fields was only a few fractions of a second, but as the field became higher it increased to several minutes. This fact, no doubt, testifies to electrically assisted aggregation of particles and permits the conclusion to be made that the size of an aggregate is about one order of magnitude larger than the size of a single particle.

When an electric field is applied to a dispersion, a situation may arise in which the relaxation time of induced conductivity anisotropy does not change but the aggregation does occur. In this case the aggregation is evidenced by changes in the conductivity anisotropy versus the time of the field action at a given value of the field strength.

It has been shown experimentally that a longer time of the field action results in that the $\delta K_{\parallel,\perp}/\bar{K}$ ratio first reaches an intermediate value corresponding to the steady-state orientation of non-interacting particles, then, some time later, it begins to grow again owing to the particle aggregation. When the aggregation process is in equilibrium with disintegration of the aggregates, the above ratio reaches another steady-state value. In this case the relaxation time of conductivity anisotropy is independent of the duration of the field action because the particles do not interact after the field turn-off. Figure 6.3 gives the dependences of $\delta K_{\parallel}/(\delta K_{\parallel})_{\max}$ on the field action time, and Figure 6.4 shows the initial time $T$ of the second steady-state as a function of $E^{-2}$.

The above processes induced by an applied electric field are reversible because they do not change the properties of the dispersion. This behavior of a dispersion in an electric field is possible up to a certain energy of particle interaction exceeding the thermal energy of Brownian motion. After the field turn-off, the interaction energy decreases abruptly, and the aggregated and orientationally ordered particles start to rotate randomly due to the thermal motion. The relaxation time of induced conductivity anisotropy did not change if the aggregates were spaced at a distance larger than the particle size.

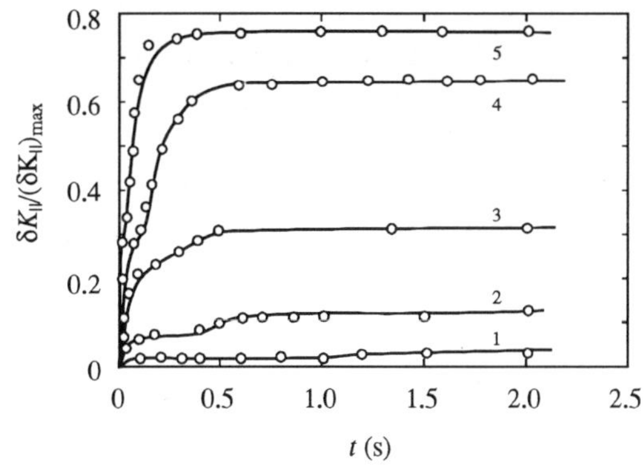


**Figure 6.3.** The plots of $\delta K_{||}/(\delta K_{||})_{\max}$ versus $t$ for a palygorskit aqueous colloidal system at different electric field strengths: 45 V cm$^{-1}$ (curve 1), 100 V cm$^{-1}$ (curve 2), 160 V cm$^{-1}$ (curve 3), 245 V cm$^{-1}$ (curve 4), 320 V cm$^{-1}$ (curve 5).

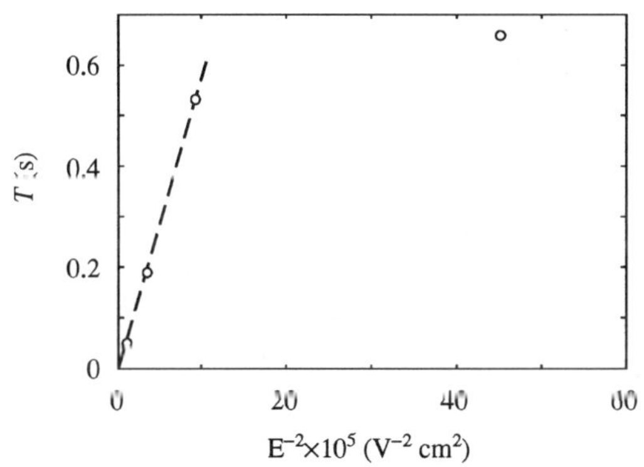


**Figure 6.4.** The plot of $T$ versus $E^{-2}$ for a palygorskit aqueous colloidal system.

To conclude, the problems of aggregation and electrically assisted aggregation of particles in dispersions are still poorly understood, both theoretically and experimentally. Some preliminary studies have shown, however, that orientational conductometry opens up perspectives for investigation of particle interactions in various dispersions.

## 6.4. EXPERIMENTAL MEASUREMENT OF ELECTRICAL CONDUCTIVITY ANISOTROPY IN DISPERSIONS

In order to measure the anisotropy of electrical conductivity of dispersions, the experimental cell is connected to a current source, which induces a sine-shaped measuring current with a time-independent amplitude. For the region to be measured we have

$$\boldsymbol{E}_{\text{meas}} = \hat{K}^{-1}\boldsymbol{j},$$

where $\boldsymbol{j}$ is the current density, $\hat{K}$ is the conductivity matrix of the dispersion.

If the electric field is directed along the $z$-axis in a system of the $x, y$ and $z$ coordinates related to the cell, the matrix $\hat{K}$ has the form:

$$\hat{K} = \begin{pmatrix} K_x & 0 & 0 \\ 0 & K_y & 0 \\ 0 & 0 & K_z \end{pmatrix},$$

where $K_x = K_y$. The components of the measuring field strength are defined as follows:

$$E_{\text{meas},i} = \frac{1}{K_i} j_i,$$

where $i = x, y, z$.

When the particles change their orientation, the $K_i$ components also change, but if $j_i$ remains constant and the relative changes of $K_i$ are small, then

$$\frac{\delta E_{\text{meas}}}{E_{\text{meas}}} = -\frac{\delta K_i}{\bar{K}}. \tag{6.19}$$

Here $\delta E_{\text{meas}}$ are the changes in the measuring field when the particles become oriented.

Experimental measurements of $\delta K_{\|,\perp}/\bar{K}$ can be made in a cell containing four probes and four electrodes which represent long thin cylinders arranged parallel to each other, as is shown schematically in Figure 6.5. An orienting field is applied to electrodes $A$ and $B$, the other two, $C$ and $D$, are the measuring electrodes. The lines $AB$ and $CD$ connecting the electrodes lie at an angle of 45° to each other. This is necessary because the measuring current density in the region of probes 1–4 should have identical projections on the lines connecting probes 1 and 3, and also 2 and 4. The distance between the probes is much less than between the electrodes, and the orienting and measuring fields in this area are actually uniform. Probes 1 and 3 can measure the relative variation in the electrical conductivity along the orienting field, while probes 2 and 4 measure the same parameter in the direction perpendicular to this field. In fact, since the measuring field in the probe area is uniform, the relative variation in the voltage $U_{\|}$ on probes 1 and 2 is equal to the relative variation of the field:

$$\frac{\delta U_{\|}}{U_{\|}} = \frac{\delta E_{\text{meas},z}}{E_{\text{meas},z}}.$$

With (6.19) we can write

$$\frac{\delta U_{\|}}{U_{\|}} = -\frac{\delta K_{\|}}{\bar{K}}.$$

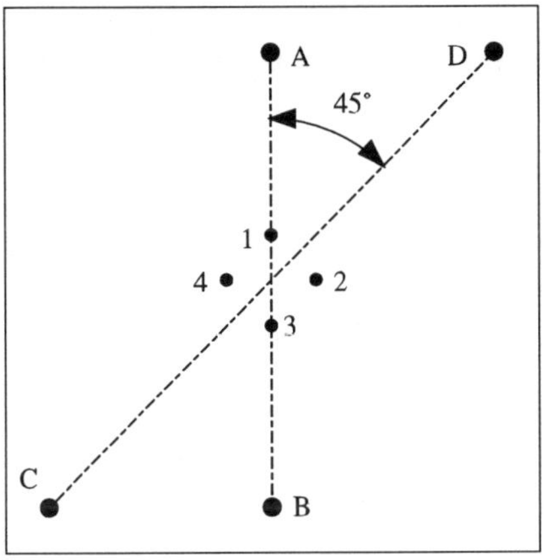


**Figure 6.5.** The arrangement of electrodes and probes in the experimental cell (cross section is normal to the electrode axis).

Similarly, for the variation in the voltage $U_\perp$ on probes 2 and 4 we have

$$\frac{\delta U_\perp}{U_\perp} = -\frac{\delta K_\perp}{\bar{K}}.$$

The measurements of $U_\parallel$ and $U_\perp$ are made by balanced and bridge circuits which provide zero-currents through the probe circuits under the equilibrium conditions corresponding to disoriented particles. When the circuit is imbalanced due to the field-induced orientation, the imbalance signal must be proportional to $U_{\parallel,\perp}$. This measuring procedure allows one to avoid the probe polarization and, thus, the error it entails. In order to avoid the influence of the orienting field in conductivity measurements, the steady-state effect is detected immediately after this orienting field is turned off while the particles are still orientationally ordered. The measurements of $\delta U_{\parallel,\perp}$ at the subsequent moments provide the relaxation times of the conductivity anisotropy.

If one measures only $\delta K_\parallel/\bar{K}$ while $\delta K_\perp/\bar{K}$ is derived from formula (6.14), one can use a more simple procedure. Platinum or other metallic electrode plates of the experimental cell are connected to the orienting field source with an electron switch or a fast-response relay, but then, at the moment of the field turn-off, they are connected to the measuring circuit input. When the electrical conductivity of a dispersion is high, the orienting field pulse will make it still higher owing to the Joule heat, which is released when the current passes through the dispersion. These changes may be comparable with the values of $\delta K_\parallel$ and $\delta K_\perp$. However, the measurement errors for both parameters will be identical, and the difference between them is described by an expression

$$\delta K = \delta K_\parallel - \delta K_\perp = \frac{4\pi n_0}{\epsilon}\bar{K}\gamma\mathcal{F}(\chi,0)$$

similar to (6.12) and (6.13).

When a cell has no probes and can measure only $\delta K_{\|}/\bar{K}$, it is necessary to eliminate the heat effect on the conductivity or to make allowance for it. One should note that the temperature increased by the orienting field will exponentially decrease after the field turn-off in much the same way as the observable anisotropy. In order to differentiate between the heat variation of conductivity and its orientational change, the relaxation times of these two processes must be essentially different. A strict temperature control of the cell and its electrodes[75] leads to an essential decrease in the thermal relaxation time, making practically undistinguishable the thermal and orientational growth of the electrical conductivity. So it is reasonable in many cases to use a cell with poor heat conductivity to distinguish between the relaxations of the thermal and orientational effects. The pulsed orienting field must have a rather long time interval between the pulses so that the dispersion temperature could return to its initial value by the moment the next orienting pulse is applied.

## 6.5. DOUBLE LAYER CHARACTERISTICS AND CONDUCTIVITY ANISOTROPY

In Section 6.2 we considered the expressions for a steady-state effect associated with polarizability anisotropy and for the relaxation of conductivity anisotropy. These expressions have very much in common with the formulas for a steady-state electrooptic effect in uniform sine-shaped electric fields. One can conclude from this that conductivity anisotropy measurements can be used to study the anisotropy of particle polarizability in a way similar to electrooptic experiments. Indeed, from (6.15) and (6.13) we can get the dependence of the conductivity anisotropy on the orienting field:

$$\frac{\delta K_{\|}(E^2)}{\bar{K}} = \frac{8\pi n_0}{3\epsilon} \int_0^{\infty} \mathcal{F}(\chi, 0)\gamma(r)\varphi(r)\mathrm{d}r \tag{6.20}$$

and from a comparison of the experimental and theoretical dependences we can find $Z = \kappa^{\sigma}/K$. The function $f(r) = \gamma(r)\varphi(r)$ necessary for these calculations can be found from the relaxation behavior of the observable effect after complete particle orientation. The equation for the function $f(r)$ can be represented as:

$$\frac{\delta K_{\|}(t)}{\bar{K}} = \frac{8\pi n_0}{3\epsilon} \int_0^{\infty} \exp[-6D(r)t]f(r)\mathrm{d}r. \tag{6.21}$$

The algorithm of experimental data processing may be the same as for electrooptic studies described in Chapter 4. If the value of $\delta K_{\|}/\bar{K}$ corresponding to a complete particle orientation along the field is known, the factor $8\pi n_0/3\epsilon$ can be eliminated by normalizing the observable effect by unity, as was done in the electrooptic study. If the number density of particles $n_0$ is known, one can calculate the absolute values of $\delta K_{\parallel}/\bar{K}$ and their variation. This permits a more detailed comparison of the theoretical and experimental dependences of $\delta K_{\parallel}/\bar{K}$ on $E^2$ than of $N$ on $E^2$. Knowledge of the absolute values of $\delta K_{\parallel}/\bar{K}$ is especially important in the study of dispersions that do not permit measurements in strong electric fields inducing a complete orientation.

For low fields, expression (6.20) can be considerably simplified, if the function $\mathcal{F}(\chi, 0)$ is expanded into series over $E^2$ (see (2.18) and (2.19)) only with the first expansion term taken into account:

$$\frac{\delta K_{\parallel}(E^2)}{\bar{K}} = \frac{8\pi n_0}{45kT\epsilon} E^2 \int_0^{\infty} \gamma(r) f(r) \mathrm{d}r. \tag{6.22}$$

When the exact size distribution function $\varphi(r)$ is known, it is expedient to use the expression

$$\frac{\delta K_{\parallel}(E^2)}{\bar{K}} = \frac{8\pi n_0}{45kT\epsilon} E^2 \int_0^{\infty} \gamma^2(r) \varphi(r) \mathrm{d}r. \tag{6.23}$$

instead of equation (6.22).

The $E^2$ dependences of $\delta K_{\perp}/\bar{K}$ can be easily derived from equations (6.22), (6.23) and (6.14).

Equations of (6.20) allow one to determine the value of $Z = \kappa^{\sigma}/K$ entering the expression for $(\gamma_{\parallel} - \gamma_{\perp})$ and to calculate the surface conductivity of particles, utilizing the experimental $E^2$ dependences of $\delta K_{\parallel,\perp}/\bar{K}$. The way of finding $Z$ in this case is the same as for the electrooptic effect due to dichroism. However, it seems more promising for the study of the problem in question.

With the anisotropy of surface conductivity, we can estimate the values of $\kappa_{\parallel}^{\sigma}$ and $\kappa_{\perp}^{\sigma}$ along and across the long axis of a particle, using the values of $\delta K_{\parallel}/\bar{K}$ and $\delta K_{\perp}/\bar{K}$ related to a complete orientation. For this we may employ the following relationships

$$\frac{(\delta K_{\parallel})_{\max}}{\bar{K}} - \frac{\Delta K}{\bar{K}} = \frac{8\pi n_0}{3\epsilon} \int_0^{\infty} \gamma_{\parallel}(Z_{\parallel}, r) \varphi(r) \mathrm{d}r$$

$$\frac{(\delta K_{\perp})_{\max}}{\bar{K}} - \frac{\Delta K}{\bar{K}} = -\frac{4\pi n_0}{3\epsilon} \int_0^{\infty} \gamma_{\perp}(Z_{\perp}, r) \varphi(r) \mathrm{d}r,$$

which can be easily derived from equations (6.8) and (6.20). The field dependences of $\delta K_{\parallel}/K$ and $\delta K_{\perp}/K$ are used to find a mean value of $Z$ lying between $Z_{\parallel}$ and $Z_{\perp}$. For a complete orientation, we can write

$$\alpha_{\text{con}}(E^2) = \frac{\delta K_{\parallel,\perp}}{(\delta K_{\parallel,\perp})_{\max}} = \int_0^{\infty} \gamma \mathcal{F}(\chi, 0) \varphi(r) \mathrm{d}r. \tag{6.24}$$

The relaxation $\alpha_{\text{con}}$ after the field turn-off is expressed as

$$\alpha_{\text{con}}(t) = \int_0^{\infty} \exp[-D(r)t] \gamma \varphi(r) \mathrm{d}r.$$

This formula permits calculation of $f(r) = \gamma\varphi(r)$ if the function $\alpha_{\text{con}}(t)$ is known.

## 6.6. EFFECT OF THE MEDIUM IONIC COMPOSITION ON ELECTRICAL CHARACTERISTICS OF PARTICLES

In this section we will try to demonstrate the advantages of orientational conductometry combined with electrokinetic measurements in the study of electrical characteristics of particles and the influence of the medium ionic composition on them. For such investigations we have chosen aqueous suspensions of palygorskit and quartz with addition of $BaCl_2$, KCl and $AlCl_3$ electrolytes. Potassium, barium and aluminum forms of Cherkask palygorskit (clay variety) were used. The procedure for their preparation was described elsewhere.[35,78] Electron microscopic measurements have shown that particles this clay variety are rods of 0.5–1.0 $\mu$m long and about 0.04 $\mu$m in diameter, with the mean semiaxis ratio $p \simeq 25$. The density of palygorskit is 2.3–2.7 g cm$^{-3}$.

In order to choose a palygorskit number density in a suspension suitable for measurements, we studied the orientational conductivity as a function of the number density for the orienting field action of 25 s. The ionic composition of the dispersions was controlled by their electrical conductivity and maintained the same for all the number densities by adding the dispersive medium separated by centrifugation from a highly concentrated palygorskit suspension.

The function $\alpha_{\text{con}}(E^2)$ was the same at the particle content of $5.0 \times 10^{-4}$ g cm$^{-3}$ and lower, the lowest content being $5.0 \times 10^{-2}$ g cm$^{-3}$ (Figure 6.6a). In this case the relaxation behavior of orientational conductivity did not change, which indicates the absence of interaction between the particles during the measurements (Figure 6.6a). For this reason, we used the particle content of $1.0 \times 10^{-4}$ g cm$^{-3}$, which was suitable for both orientational conductivity and microelectrophoretic measurements.

The relative experimental dependences of orientational conductivity $\alpha_{\text{con}}(E^2)$ at low electrolyte concentrations are shown in Figure 6.7. The relaxation behavior of conductivity was practically the same for all the electrolytes. The value of $\alpha_{\text{con}}(E^2)$ decreased by a factor of 2.7 during a period of about 0.05 s (see Figure 6.6). Therefore, the duration of the orienting field action of 0.25 s was sufficient for the steady-state orientation of particles to be established. The curve for $\alpha_{\text{con}}(E^2)$ is shown in Figure 6.6a at zero electrolyte concentration in a palygorskit aqueous suspension. Figure 6.6b gives the size distribution functions $f(r) = \gamma\varphi(r)$ and $\varphi(r)$ with the polarization weight and without it found from $\alpha_{\text{con}}(E^2)$ and $\alpha_{\text{con}}(t)$.

In Figure 6.7 the experimental values of $\alpha_{\text{con}}(E^2)$ are shown with dotted lines, and the values calculated from equation (6.24) are represented with solid lines. We have used the polarizability model for ellipsoids of revolution. One can see a good agreement between experiment and theory, which testifies to the applicability of this model for description of the orientational conductometric effect in this particular case.

The orienting field frequency of 5 kHz in our experiments corresponded to a plateau in the frequency dependence of conductivity and of electroop-

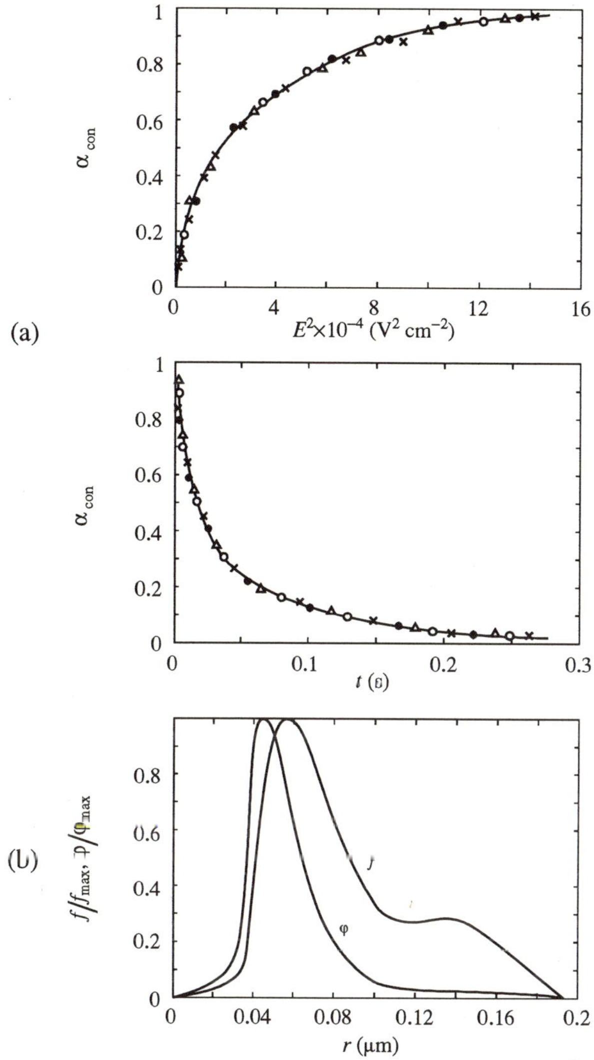


**Figure 6.6.** (a) The experimental and theoretical dependences of $\alpha_{con}$ for palygorskit aqueous suspensions at various particle contents. ○, $5.0 \times 10^{-4}$ g cm$^{-3}$; △, $2.5 \times 10^{-4}$ g cm$^{-3}$; ×, $1.0 \times 10^{-4}$ g cm$^{-3}$; ●, $5.0 \times 10^{-2}$ g cm$^{-3}$. (b) The theoretical plots of $f/f_{max}$ and $\varphi/\varphi_{max}$ obtained from the data presented in (a).

tic effects associated with light scattering and dichroism in the suspensions. An electrooptic study of the suspensions showed a good theoretical accuracy in the description of orientational effects within the polarizability model for ellipsoids of revolution. Orientational conductometry has demonstrated the field frequency of 5 kHz to be far from the region of frequency-dependent conductivity not only for palygorskit systems but also for all the other colloidal

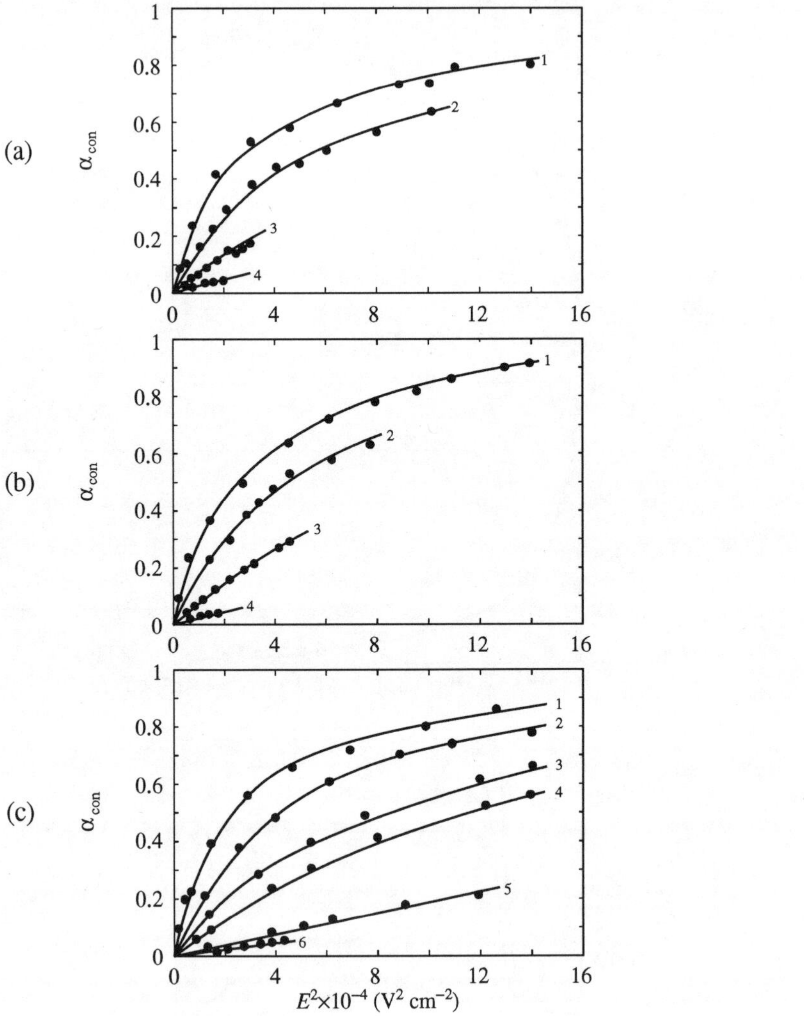


**Figure 6.7.** The plots of $\alpha_{con}(E^2)$ for palygorskit aqueous suspensions at various concentrations of the electrolytes. (a) KCl: $3.0 \times 10^{-5}$ mol dm$^{-3}$ (curve 1), $1.0 \times 10^{-4}$ mol dm$^{-3}$ (curve 2), $3.0 \times 10^{-4}$ mol dm$^{-3}$ (curve 3), $1.0 \times 10^{-3}$ mol dm$^{-3}$(curve 4). (b) $BaCl_2$: $3.0 \times 10^{-6}$ mol dm$^{-3}$ (curve 1), $3.0 \times 10^{-5}$ mol dm$^{-3}$ (curve 2), $1.0 \times 10^{-4}$ mol dm$^{-3}$ (curve 3), $3.0 \times 10^{-4}$ mol dm$^{-3}$ (curve 4). (c) $AlCl_3$: $1.0 \times 10^{-6}$ mol dm$^{-3}$ (curve 1), $3.0 \times 10^{-6}$ mol dm$^{-3}$ (curve 2), $1.0 \times 10^{-5}$ mol dm$^{-3}$ (curve 3), $3.0 \times 10^{-5}$ mol dm$^{-3}$ (curve 4), $1.0 \times 10^{-4}$ mol dm$^{-3}$ (curve 5), $3.0 \times 10^{-4}$ mol dm$^{-3}$ (curve 6).

suspensions studied. This result is supported by the electrooptic experiments described in Chapter 4. For this reason, this orienting field frequency was used in further measurements.

We also measured $K$ and $Z$ in order to find the surface conductivity $\kappa^\sigma$ of palygorskit particles at the chosen electrolyte concentrations. The relative

(a)

(b)

(c)

(d)

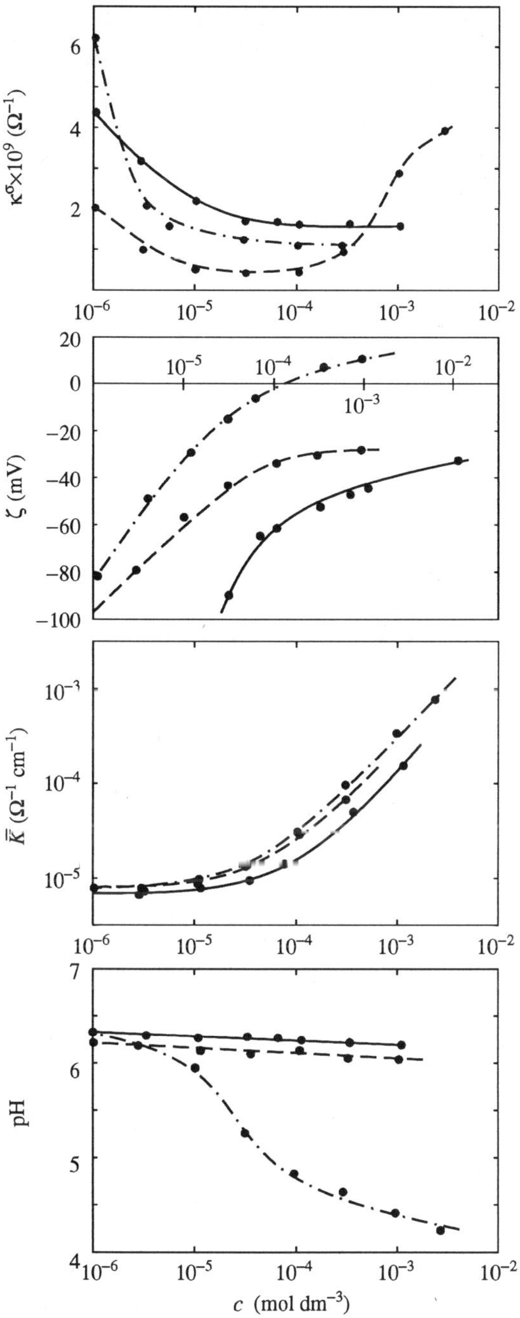


**Figure 6.8.** A palygorskit aqueous suspension with an added electrolyte. (a) The plots of $\kappa^\sigma$ versus the electrolyte concentrations. KCl (solid curve), $BaCl_2$ (dashed curve), $AlCl_3$ (dash-dotted curve). (b) The same for $\zeta$. (c) The same for $\bar{K}$. (d) The same for pH.

**Table 6.1**

| | | $\kappa^\sigma \times 10^9, \Omega^{-1}$ | |
|---|---|---|---|
| Electrolyte | $c$, mol dm$^{-3}$ | Present data | Data[39] |
| NaCl | $1.0 \times 10^{-4}$ | 4.0 | 5.7 |
| | $2.0 \times 10^{-4}$ | 4.0 | |
| | $3.0 \times 10^{-4}$ | 3.9 | |
| | $4.0 \times 10^{-4}$ | 3.8 | |
| | $7.0 \times 10^{-4}$ | 3.6 | 5.7 |
| | $3.9 \times 10^{-3}$ | 3.6 | 1.6 |
| KCl | 0.0 | 4.1 | |
| | $3.0 \times 10^{-6}$ | 1.7 | |
| | $5.8 \times 10^{-6}$ | | 2.4 |
| | $6.0 \times 10^{-5}$ | 2.7 | |
| | $1.0 \times 10^{-4}$ | 1.6 | |
| | $2.6 \times 10^{-4}$ | | 2.1 |
| | $3.0 \times 10^{-4}$ | 1.6 | |
| | $1.0 \times 10^{-3}$ | 1.6 | |
| $BaCl_2$ | $3.0 \times 10^{-6}$ | 2.0 | |
| | $1.0 \times 10^{-5}$ | 1.5 | |
| | $3.0 \times 10^{-5}$ | 1.3 | |
| | $1.0 \times 10^{-4}$ | 1.2 | |
| | $3.0 \times 10^{-4}$ | 1.2 | |
| $AlCl_3$ | $2.7 \times 10^{-5}$ | | 2.3 |
| | $3.0 \times 10^{-5}$ | 0.5 | |
| | $5.0 \times 10^{-5}$ | | 1.5 |
| | $1.0 \times 10^{-4}$ | 0.4 | |
| | $3.0 \times 10^{-4}$ | 1.0 | |
| | $4.2 \times 10^{-4}$ | | 0.15 |
| | $1.0 \times 10^{-3}$ | 2.9 | |
| | $3.0 \times 10^{-3}$ | 4.0 | |

error in $\kappa^\sigma$ was found from the model function computations to be about 3%. The ionic composition of the systems was controlled by measuring pH of the medium after the electrolytes were added to the suspensions. Besides, the electrophoretic mobility of particles was determined by microelectrokinetic techniques at each electrolyte concentration. Since there is no rigorous theory to describe the electrophoretic mobility of strongly elongated rods of such a small size at low electrolyte concentrations, the $\zeta$-potential of palygorskit particles was estimated by Smolukhovskii's formula[48] (see equation (3.13)).

The values of $\kappa^\sigma$, $\zeta$, $\bar{K}$ and pH against the electrolyte concentrations are given in Figure 6.8. One can see that there is a certain correlation between the concentration dependences of $\kappa^\sigma$ and $\zeta$ for all the electrolytes. The value of $\kappa^\sigma$ is practically unaffected by the concentration of KCl in the suspension starting with a concentration of $6 \times 10^{-5}$ mol dm$^{-3}$. As to $BaCl_2$, this value of the concentration is lower than for KCl. As the $AlCl_3$ concentration grows beyond

$10^{-6}$ mol dm$^{-3}$, the value of $\kappa^\sigma$ drops down to $0.44 \times 10^{-9}\ \Omega^{-1}$ and then begins to grow. This behavior of $\kappa^\sigma$ on addition of electrolyte is consistent with our understanding of the properties of an electrical double layer in dilute electrolyte solutions.[45]

The hydrate envelopes of alkali metals are not destroyed in their interaction with the particle surface, so that the minimum distance between an alkali cation and the surface corresponds to the cation localization in the external Helmholtz plane. Counterions easily migrate along the surface, however their mobility may be somewhat lower than in the electrolyte bulk. With increasing electrolyte concentration the diffuse double layer shrinks, but the fraction of cations localized in the external Helmholtz plane becomes large only at sufficiently high concentrations of the electrolytes,[58] e.g. over $10^{-3}$ mol dm$^{-3}$. Due to the stronger interaction with the particle surface, the mobility of multivalent ions turns out to be lower than in the bulk, even in dilute electrolytic solutions.

A sharp decrease in $\kappa^\sigma$ for the aluminum variety of palygorskit, as the $AlCl_3$ concentration grows, may be due not only to the lower ion mobility at the particle surface but to the smaller charge of the diffuse double layer because of the adsorption of $Al^{+3}$ ions. In the $AlCl_3$ electrolyte, the sign reversal of the charge occurs on the particle surface. The value of $\kappa^\sigma$ drops to its minimum in the charge-reversal region but it does not vanish. This can be accounted for by the fact that mobile charges on the surface do not disappear entirely when the surface charge changes its sign, but they are compensated by the opposite charges. Another reason is that the electrical and geometrical parameters of the surface vary with the particle species. The available data on the behavior of aluminum ions in an aqueous solution and on the composition of structures formed in hydrolysis as a function of the medium concentration and pH have been systematized in a review.[79] The following ions are present in solutions of aluminum salts at the concentrations $10^{-6}$–$10^{-4}$ mol dm$^{-3}$ and in the pH range from 4.2 to 6.0:

$$Al^{+3}, \quad [Al(OH)(H_2O)_5]^{+2}, \quad [Al(OH)_2(H_2O)_4]^{-1}.$$

According to E. Matijevich,[80,81] aluminum hydrocomplexes possess such a high adsorptivity that they can cause a charge reversal on colloidal particles, and this is what we observed in our experiments (Figure 6.8).

In order to prevent particle coagulation, the dispersions were prepared immediately prior to the experiment by adding a definite quantity of the material to an $AlCl_3$ solution of a given concentration. The $AlCl_3$ solutions were kept for 24 hours for equilibrium hydrolysis to be established.

The surface conductivity $\kappa^\sigma$ of palygorskit in aqueous suspensions has been studied by several researchers.[58,82,83] Table 6.1 presents $\kappa^\sigma$ data obtained from orientational conductometry, electric dichroism studies, and from measurements of $\bar{K}$ and $K$.

The differences in the $\kappa^\sigma$ values obtained by different authors for the same electrolyte may be due to the following reasons: variation in the electrolyte composition because of the different preparation procedures used (for example, $AlCl_3$ solutions were not aged for 24 hours prior to measurements[83]), different sources of the mineral and different treatment procedures, and the use of a monodisperse approximation[82,83] in the calculations of a polydisperse system with a number density 100 times higher than in our measurements.

To estimate the ion mobility within the double layer of a palygorskit particle, we will use a formula from:[45]

$$u^+ = \frac{0.7\mathrm{Rel}}{\tau(1+3k_m)},$$

where $\tau$ is the ion exchange volume[35,45] which is assumed to be $1.1 \times 10^{-4}$ g-equiv./g; $k_m$ is a parameter characterizing the contribution to the conductivity by the charge transfer via the electroosmotic flow of the fluid, which is equal to 0.186 for KCl.[45] For estimation of the half-thickness $b$ of palygorskit particles necessary for the calculation of $\mathrm{Rel} = \kappa^\sigma/Kb$, we will use the size distribution function of particles for spheres equivalent in volume to the particles. One can see from Figure 6.6 that the most probable radius of a sphere equivalent in volume to a particle is $r = 0.41$ $\mu$m. The value of $b$ can be found by the formula

$$b = rp^{-1/3}, \tag{6.25}$$

where $p = a/b = 25$, $a$ is the long semiaxis equal to 0.35 $\mu$m in this case. Formula (6.25) follows from the equality of volumes of a sphere and an ellipsoid: $4\pi r^3/3 = 4\pi ab^2/3$.

Figure 6.7 demonstrates the coincidence of the shapes of the theoretical and experimental dependences of $\alpha_{\mathrm{con}}$ on $E$ for a palygorskit suspension. However, it is interesting to quantitatively compare the experimental and theoretical values for $\delta K_\parallel/K$. To estimate this value, it is necessary to find the particle number density in the system, so we will consider the function $\varphi(r)$. The volume of a particle averaged over a statistical ensemble follows from the expression

$$\bar{V} = \frac{4\pi}{3}\int_{r_{\min}}^{r_{\max}} \varphi(r) r^3 \mathrm{d}r = 1.0 \times 10^{-3}\ \mu\mathrm{m}.$$

Then, with allowance for the palygorskit density $\rho = 2.7$ g cm$^{-3}$,[84] we obtain for the mean mass of a single particle $\bar{m} = \bar{V}_\rho = 2.7 \times 10^{-15}$ g. Keeping in mind that the particle content in the colloidal solution is $1.0 \times 10^{-4}$ g cm$^{-3}$, we find the number density $n_0 = 3.7 \times 10^{10}$ cm$^{-3}$. According to the expression for a conductometric effect, at $E \to \infty$ it is possible to write

$$\frac{\delta K_\parallel}{\bar{K}} = -\frac{\delta U_\parallel}{U} = \frac{8\pi n_0}{3\epsilon}\int_{r_{\min}}^{r_{\max}} \gamma(r)\varphi(r)\mathrm{d}r.$$

For intermediate values one should make use of formula (6.20).

**Table 6.2**

| Electrolyte | $c$ mol dm$^{-3}$ | $\bar{\gamma} \times 10^{13}$, cm$^3$ | $\frac{\delta K_{\parallel}}{\bar{K}} \times 10^4$ Calculation | Experiment |
|---|---|---|---|---|
| KCl | $1.0 \times 10^{-5}$ | 2.5 | 1.2 | 0.90 |
| | $3.0 \times 10^{-5}$ | 2.1 | 1.0 | 0.80 |
| | $6.0 \times 10^{-5}$ | 1.5 | 0.73 | 0.63 |
| | $1.0 \times 10^{-4}$ | 1.3 | 0.61 | 0.50 |
| $BaCl_2$ | $3.0 \times 10^{-5}$ | 2.5 | 1.2 | 0.90 |
| | $1.0 \times 10^{-5}$ | 2.2 | 1.0 | 0.83 |
| | $3.0 \times 10^{-5}$ | 1.4 | 0.73 | 0.62 |
| | $1.0 \times 10^{-4}$ | 0.60 | 0.39 | 0.30 |
| $AlCl_3$ | $1.0 \times 10^{-6}$ | 2.4 | 1.2 | 0.86 |
| | $3.0 \times 10^{-6}$ | 1.6 | 0.77 | 0.75 |
| | $1.0 \times 10^{-5}$ | 1.3 | 0.66 | 0.65 |
| | $3.0 \times 10^{-5}$ | 1.1 | 0.56 | 0.55 |
| | $1.0 \times 10^{-4}$ | 0.40 | 0.20 | 0.20 |

Table 6.2 gives the calculated and measured values of $\delta K_{\parallel}/\bar{K}$ and $\bar{\gamma}$. It is seen that the theoretical results exceed the experimental ones by about 20%. This fact can be explained as follows. Palygorskit particles were treated by ultrasound (22 kHz) for 5 min and then studied under an electron microscope. The observations showed that only 80% aggregates had decomposed after the treatment. If we assume that in the remaining aggregates the particles were connected by their lateral surfaces along their entire length where the binding energy is maximum, the aggregates will possess the same rotational diffusion constant as the individual particles because of their great length ($p = 25$). If we compare only the shapes of the curves for the field dependence of orientational conductivity and of its relaxation behavior after the field turn-off the value of $\gamma$ for the aggregates will not differ from $\gamma$ for single particles. Particle aggregation affects only the value of the orientational effect, which can be detected by orientational conductometry.

A quartz dispersion is a very convenient model object for a quantitative study of the double layer characteristics. Quartz particles in a colloidal system have a shape close to spherical, which permits application nearly any of the available theories of electrophoresis.[48] On the other hand, quarts particles differ from a sphere, so that this difference makes the use of orientational methods also profitable. Moreover, the size of quartz particles of 1–2 $\mu$m allows microscopic observations to be made at the instrument resolution of 0.2–0.3 $\mu$m in order to control the particle sizes and shapes.

In our study we used an aqueous suspension of amorphous quartz prepared by the procedure.[85] Besides, the quartz powder was mixed with distilled water several times in the proportion 1:20 and allowed to precipitate for 24 hours. For experiment we used only the portion of quartz which did not precipitate

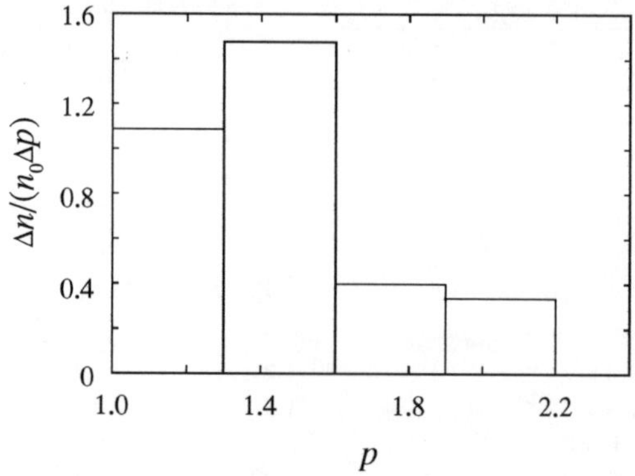


**Figure 6.9.** A histogram of quartz particle distribution.

during this period. This quartz dispersion was centrifuged, dried and weighed with an analytical balance in order to determine the particle concentration in the suspension. Before measurements the quartz was brought in equilibrium with the electrolyte for four hours and, immediately before the experiment, it was treated with ultrasound of 22 kHz.

The electrical and geometric characteristics of amorphous quartz were studied using electro-orientational conductometry, electric dichroism and microelectrophoresis. The quartz concentration in the suspension was $1.0 \times 10^{-4}$ g cm$^{-3}$ for electrooptic, electrophoretic and conductivity measurements, because the values of $\alpha_{\text{con}}(E^2)$, $\alpha_{\text{con}}(t)$, $N(E^2)$ and $N(t)$ did not change at a concentration below $5.0 \times 10^{-4}$ g cm$^{-3}$. The orienting field frequency for conductometric and electrooptic studies was taken to be 5 kHz, which corresponds to a plateau on the respective curves. The orienting pulse duration was 0.5 s, which was sufficient for a steady-state particle orientation at any field strength.

From a microscopic study we obtained a histogram of particle distribution with respect to the semiaxis ratio $p$ (Figure 6.9). One can notice that the values of $p$ in the range from 1.3 to 1.6 have the highest probability. In the analytical description of the orientational conductivity function $\alpha_{\text{con}}(E^2)$, the best fit between the theoretical and experimental curves was observed at $p = 1.4$, which is consistent with the data obtained by light microscopy. The value $p = 1.4$ was taken for the calculation of $Z$ from the electrooptic and conductometric data. The time dependence of the conductometric effect, $\alpha_{\text{con}}(t)$, and the function $f(r)$ calculated from $\alpha_{\text{con}}(t)$ are given in Figure 6.10 without electrolyte addition to the quartz suspension. The solid line stands for the calculated $\alpha_{\text{con}}(E^2)$, and one can also see the function $\varphi(r)$ derived from the experimental $\alpha_{\text{con}}(E^2)$. The theoretical and calculated curves for $\alpha_{\text{con}}(E^2)$ fit each other fairly well.

The methods of orientational conductometry and electrical dichroism were

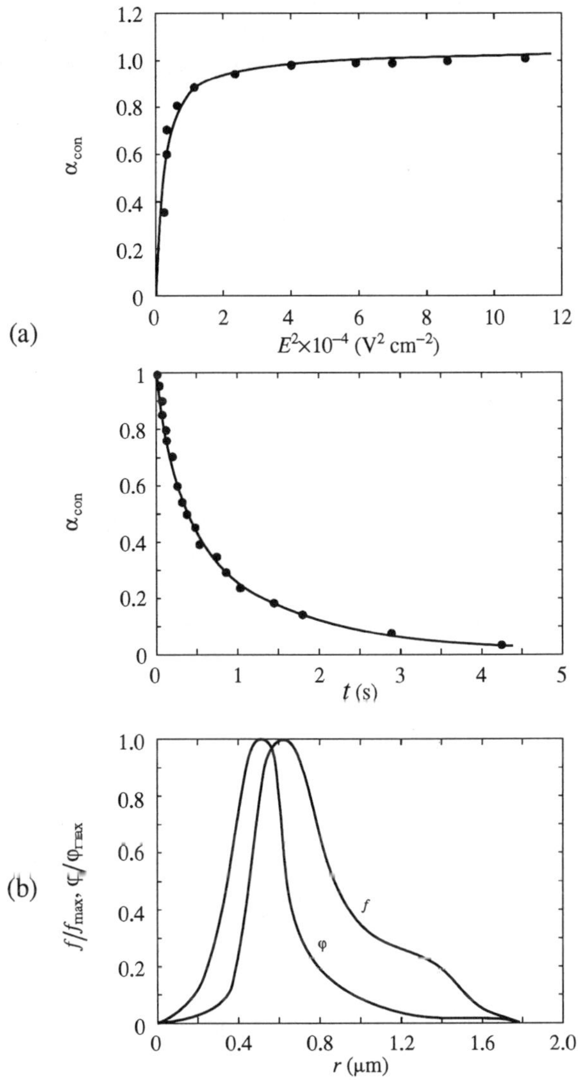


**Figure 6.10.** (a) The experimental and theoretical dependences of $\alpha_{con}$ for a quartz aqueous suspension. (b) The theoretical plots of $f/f_{max}$ and $\varphi/\varphi_{max}$ obtained from the data presented in (a).

used to study the dependences of $\kappa^{\sigma}$ for quartz particles on KCl, $BaCl_2$ and $AlCl_3$ electrolyte concentrations. Computations with model functions showed a relative error in the $\kappa^{\sigma}$ values to be 3–5% in both electrooptic and conductometric data. Together with $\kappa^{\sigma}$, we determined the $\zeta$-potential of the quartz particles, $\bar{K}$ and pH of the suspension. The respective curves for the electrolyte concentration dependences of these parameters are given in Figure 6.11. One can see an agreement between the theoretical and experimental values of $\alpha_{con}(E^2)$. This agreement and the fact that the relaxation times of the conductometric and electrooptic effects did not practically change on addition of the electrolytes indicates the absence of particle interactions during the measurements.

The electrokinetic potential of quartz particles was calculated by the Wirsema method,[86] since in the region of low electrolyte concentrations we had to take into account the relaxation effect associated with the action of the polarization charge of the diffuse double layer on the particle surface charge.[48,86] In accordance with the particle size distribution (Figure 6.12), the particle radius was taken to be 0.62 $\mu$m. The error in the $\zeta$-potential determined by the Wirsema method did not exceed 1%. The electric field in the electrophoretic cell of a closed type with reversible electrodes was several volts per centimeter.

As is seen from Figure 6.11a, there is an agreement between the values of $\kappa^\sigma$ obtained from the conductometric and electrooptic studies for all types of electrolyte over the entire range of concentrations used in the experiment. Like in case of palygorskit, the sign reversal of the quartz particle charge occurs only on addition of $AlCl_3$ (Figure 6.11b). At the charge-reversal point, $\kappa^\sigma$ does not vanish (Figure 6.11a) for the same reason as for palygorskit.

For both quartz and palygorskit, the monotonic decrease of $\kappa^\sigma$ with increasing KCl and $BaCl_2$ concentrations is in agreement with the available concepts of the double layer structure and is accounted for by ionic depletion of the layer due to the lower charge on the particle's surface.

The calculated values of orientational conductivity for a quartz aqueous dispersion with a particle density of $2.85 \times 10^7$ cm$^{-3}$ are presented in Table 6.3. The shape of the $\varphi(r)$ function necessary for the calculations is shown in Figure 6.12. It follows from Table 6.3 that there is a satisfactory agreement between the theoretical and experimental values.

The properties of quartz surface and the $\zeta$-potential have been considered in a number of papers,[48,85,87] because quartz aqueous suspensions are very convenient for model experiments.

A detailed study of $\kappa^\sigma$ and $\zeta$ for quartz as a function of KCl concentration in a suspension was done in the work.[85] The value of $\kappa^\sigma$ was calculated from the measurements of $\bar{K}$ and Rel, with the following two pairs of equations,[45] using a spherical approximation

$$\bar{K} = \frac{K}{F} + \frac{\kappa^\sigma S}{F(1-P)} \tag{6.26}$$

$$\mathrm{Rel} = \frac{\bar{K}/K + 3P/2 - 1}{3P - \bar{K}/K - 1} \tag{6.27}$$

and a spheroidal approximation

$$\bar{K} = \frac{K}{F} + \frac{2}{3} S\kappa^\sigma \frac{1 + 4b/a}{1 + \kappa^\sigma/Ka}, \tag{6.28}$$

(a)

(b)

(c)

(d)

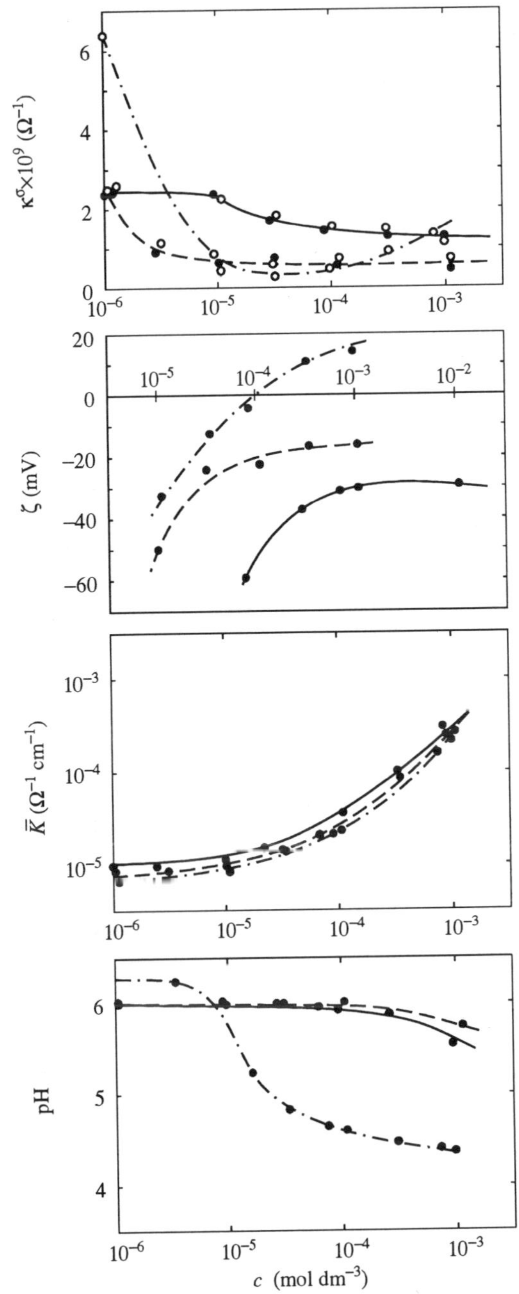


**Figure 6.11.** A palygorskit aqueous suspension with an added electrolyte. KCl (solid curves), $BaCl_2$ (dashed curves), $AlCl_3$ (dash-dotted curves). ○, electrooptical method; •, orientational conductometry. (a) The plots of $\kappa^\sigma$ versus the electrolyte concentration. (b) The same for $\zeta$. (c) The same for $\bar{K}$. (d) The same for pH.

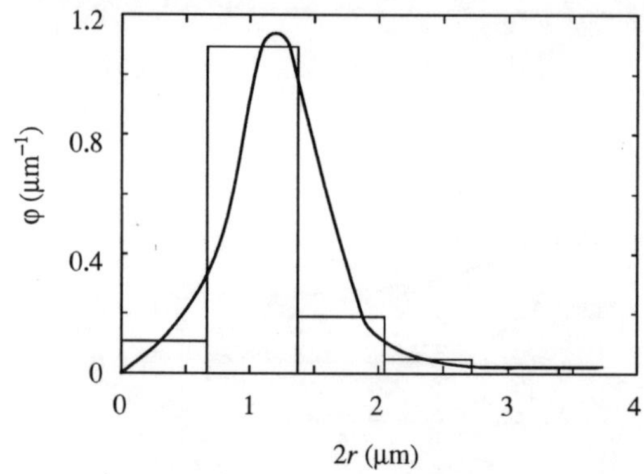


**Figure 6.12.** The plot of $\varphi(r)$ and a histogram of particle size distribution for an amorphous quartz dispersion.

**Table 6.3**

| Electrolyte | $c$ mol dm$^{-3}$ | $\bar{\gamma} \times 10^{11}$ cm$^3$ | $\frac{\delta K_{\|\max}}{K} \times 10^6$ Calculation | $\frac{\delta K_{\|\max}}{K} \times 10^6$ Experiment |
|---|---|---|---|---|
| KCl | 0.0 | 3.7 | 13 | 13 |
| | $1.0 \times 10^{-5}$ | 2.7 | 9.7 | 8.0 |
| | $3.0 \times 10^{-5}$ | 2.3 | 8.3 | 8.0 |
| | $6.0 \times 10^{-5}$ | 1.4 | 5.1 | 6.4 |
| | $1.0 \times 10^{-4}$ | 1.3 | 4.8 | 4.8 |
| | $3.0 \times 10^{-4}$ | 0.80 | 2.8 | 3.2 |
| $BaCl_2$ | $1.0 \times 10^{-6}$ | 2.7 | 9.7 | 11 |
| | $3.0 \times 10^{-6}$ | 1.8 | 6.5 | 7.0 |
| | $1.0 \times 10^{-5}$ | 1.5 | 5.3 | 6.4 |
| | $3.0 \times 10^{-5}$ | 1.3 | 4.6 | 5.3 |
| | $1.0 \times 10^{-4}$ | 0.80 | 2.7 | 2.4 |
| $AlCl_3$ | $1.0 \times 10^{-6}$ | 3.7 | 13 | 14 |
| | $1.0 \times 10^{-5}$ | 1.1 | 4.0 | 4.8 |
| | $3.0 \times 10^{-5}$ | 0.70 | 2.6 | 2.2 |
| | $6.0 \times 10^{-5}$ | 0.70 | 2.6 | 2.2 |
| | $1.0 \times 10^{-4}$ | 0.70 | 2.6 | 2.6 |

$$\text{Rel} = \frac{\bar{K}/K + 3P/2 - 1}{3P/2 - 2\bar{K}/K + 2}, \tag{6.29}$$

where $a$ is the radius of a particle, $S$ is the surface area of a particle per unit volume of the suspension, $b/a$ is a constant characterizing the deviation of the particle shape from spherical.

For an aqueous suspension of amorphous quartz Table 6.4 lists the values of $\kappa^\sigma$ obtained from orientational conductometry and electrooptic measurements

**Table 6.4**

| Concentration of KCl, mol dm$^{-3}$ | $\kappa^\sigma \times 10^9$, $\Omega^{-1}$ Orientational conductivity measurements | Electro-optical measurements | Calculation by Bickerman's formula | $-\psi_d$, mV | $-\zeta$, mV |
|---|---|---|---|---|---|
| $1.0 \times 10^{-5}$ | 2.4 | 2.3 | | 183 | |
| $3.0 \times 10^{-5}$ | 1.8 | 1.9 | | | |
| $6.0 \times 10^{-5}$ | 1.5 | 1.5 | $9.9 \times 10^{-2}$ | 167 | 69 |
| $1.0 \times 10^{-4}$ | 1.4 | 1.5 | $9.1 \times 10^{-2}$ | 172 | 60 |
| $3.0 \times 10^{-4}$ | 1.3 | 1.3 | $8.2 \times 10^{-2}$ | 149 | 44 |
| $1.0 \times 10^{-3}$ | 1.2 | 1.2 | $9.2 \times 10^{-2}$ | 108 | 35 |

**Table 6.5**

| Concentration of KCl, mol dm$^{-3}$ | $-\psi_d$, mV | | $-\zeta$, mV | | |
|---|---|---|---|---|---|
| | Data[80] | Present data | Data[53] | Data[72] | Present data |
| $1.0 \times 10^{-5}$ | 136 | 183 | 75 | 75 | |
| $5.0 \times 10^{-5}$ | 163 | | | 94 | |
| $6.0 \times 10^{-5}$ | | 167 | | | 69 |
| $1.0 \times 10^{-4}$ | 109 | 172 | 74 | 98 | 60 |
| $3.0 \times 10^{-4}$ | | 149 | | | 44 |
| $1.0 \times 10^{-3}$ | 85 | 108 | 55 | 89 | 35 |

and calculated by the Bickerman formula. One also finds here the values of $\psi_d$ and $\zeta$ which are necessary for the calculation. The values of $\zeta$ were measured experimentally, while the $\psi_d$ values were calculated by the formula[20]

$$\psi_d = -\frac{kT}{e}\ln(\kappa a \mathrm{Rel}).$$

The parameters $\zeta$ and $\psi_d$ for quartz particles in water have been determined elsewhere.[88–91] Somewhat lower values of $\zeta$ obtained in our calculations compared to the data[88,90] may be due to the fact that we studied amorphous quartz which possesses lower $\zeta$ than crystalline quartz under the same conditions.[92] However, for both kinds of quartz the Stern potential $\psi_d$ in most cases exceeds the $\zeta$-potential (Table 6.5).

As is seen from Table 6.4, there is a good agreement between the values of $\kappa^\sigma$ obtained by orientational conductometry and electrooptic methods. Slightly different values of $\kappa^\sigma$ were obtained from conductivity measurements in concentrated dispersions and quartz diaphragms.[85] The observed differences are likely to result from the monodisperse approximation employed in the calculation of $\kappa^\sigma$ by formulas (6.26) to (6.29). The measurements may also have been affected by the double layer overlapping in membranes and diaphragms with high particle densities. However, the experimental values of $\kappa^\sigma$ differ noticeably from those calculated by the Bickerman formula (see Table 6.4). The fact that, for quartz particles, the measured values of $\kappa^\sigma$ are by an order of magnitude higher than the theoretical ones and that $|\psi_\mathrm{d}| > |\zeta|$ indicates that only a small portion of the double layer ions are involved in electrokinetic phenomena.

The results presented in this section and their consistency with the respective data obtained by other methods demonstrate that orientational conductometry can be effectively used in investigations of surface conductivity of particles in polydisperse systems.

## 6.7. DETERMINATION OF POLYDISPERSITY BY THE CONDUCTOMETRIC METHOD

Let us consider the polydispersity of colloidal systems and suspensions with account of relaxation of field-induced conductivity anisotropy, as we did when discussing electrooptic effects. From the time dependence of $\delta K_{\|}(t)/\bar{K}$ in expression 6.21, we can find the function

$$f(r) = \gamma\varphi(r).$$

In Section 2.3, $\varphi(r)$ was found from analysis of the dichroic weight of $\Delta K$. Here we derive $\varphi(r)$ from the analysis of the polarizability anisotropy $\gamma$. The form of the function $\gamma(r)$ has been found without imposing restrictions on the particle size from the large values of $r$. Therefore, $\varphi(r)$ can be found in a wider range of $r$ values than in the electrooptic studies.

If we assume a particle to be an ellipsoid of revolution and approximate its polarizability anisotropy by the power dependence $\gamma \simeq cr^n$, where $c$ is a constant dependent on the shape and electrical properties of particles, the power index will lie within the limit $2.3 < n < 3.1$. When

$$\frac{\kappa^\sigma}{Kr} \geq 20 \quad \text{or} \quad \frac{\kappa^\sigma}{Kr} \leq 0.5, \tag{6.30}$$

we may suggest that the polarizability anisotropy is proportional to the particle's volume, i.e. $n \simeq 3$. Generally, this correlation is observed either in highly conductive dispersions or when the ion concentration in a system is close to zero and the double layer effect is very weak. However, when $n = 3$ and the frequency of the applied electric field is 100 kHz, $\gamma$ can be described using the Maxwell–Wagner polarizability mechanism regardless of the state of the double layer. Then the function $\varphi(r)$ has the form:

$$\varphi(r) = \frac{cf(r)}{r^3}.$$

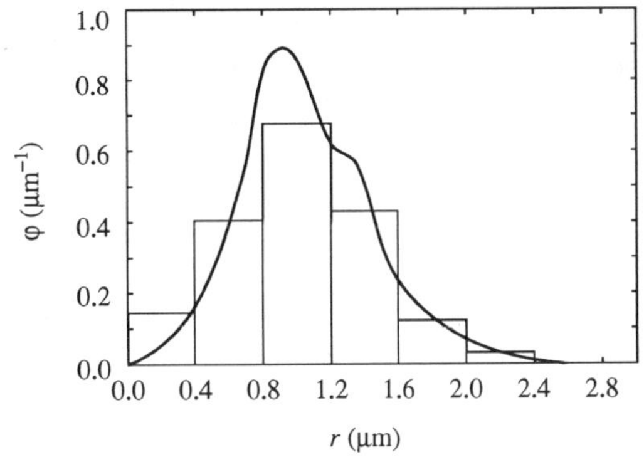


**Figure 6.13.** The plot of $\varphi(r)$ and a histogram of particle size distribution for a fused quartz aqueous suspension.

The constant $c$ can be found from the normalization condition for $\varphi(r)$. Keeping in mind that the equality

$$\int_0^\infty \varphi(r)\mathrm{d}r = 1$$

is valid for $\varphi(r)$, we determine $c$ from the following expression

$$c = \left(\int_0^\infty f(r)r^{-3}\mathrm{d}r\right)^{-1}.$$

If the frequency of the electric field is not too low and inequalities (6.30) are not fulfilled, we can describe $\gamma$ with relations (4.6) and (4.7) by approximating the particles by elongated spheroids, or with equations (4.9) and (4.10) by approximating them by flattened spheroids. One can estimate the validity of inequality (6.30) from the $E^2$ dependence of $\delta K_{\|,\perp}/\bar{K}$ and from the function $f(r)$ if one knows the parameter $Z$ and the size variation of particles within the dispersion. When approximating the particles by elongated spheroids, we can write the following expression for $\varphi(r)$

$$\varphi(r) = \frac{3f(r)}{\epsilon r^3}\left(\frac{\tau_\|}{2 + A_x\tau_\|} - \frac{\tau_\perp}{2 + (1 - A_x)\tau_\perp/2)}\right)^{-1}, \qquad (6.31)$$

where

$$\tau_\| = 3\frac{I_\| Z}{p^{2/3}r} - 2 \qquad \tau_\perp = 3\frac{I_\perp p^{1/3} Z}{r} - 2.$$

If the particles can be considered as flattened spheroids, it is easy to get an expression for $\varphi(r)$ similar to equation (6.31). For this, one should use formulas (4.9) and (4.10) instead of (4.6) and (4.7).

According to equation (6.31), to determine $\varphi(r)$ one should know not only $Z$ but the semiaxis ratio $p$ for an ellipsoid. If $p$ differs considerably from unity, then even a 2- or 3-fold change in its value will be unable to give a noticeable error in $\varphi(r)$. If $p$ is close to unity, its value can be found numerically from the normalization condition of the function $\varphi(r)$ of (6.31).

Figure 6.13 illustrates a way of finding $\varphi(r)$ (solid curve) from the studies of conductivity anisotropy for an aqueous suspension of fused quartz. This Figure also shows a histogram made on the basis of measurements with a polarization microscope. The comparison demonstrates a good agreement between the histogram and the solid line, and this again supports the reliability and validity of the information obtained from the study of electrical conductivity anisotropy in polydisperse systems.

# Conclusion

In this book we have analyzed a number of problems that can be solved by electrooptic methods and by measuring the electrical conductivity of dispersions. Calculations and experiments with real dispersions have demonstrated that polydispersity plays a decisive role in the electro-orientational mechanism of particle interaction with an applied electric field. It has been shown that particle polarizability can be determined from the data on electrical conductivity and electrooptic effects. For this it is necessary to find the relaxation behavior of dispersions and the steady-state anisotropy induced by an external field. A comparison of results on the anisotropy of electrical conductivity and electrooptic effects has shown that they provide the same or supplementary information on the dispersed material. A combination of electrooptic methods and orientational conductometry may prove very effective in the study of polyelectrolytic biological solutions and cell suspensions. Measurements of the conductivity anisotropy in polypeptide and DNA solutions can be used to calculate the polarizability of these molecules and to study their double layer. This makes it possible to determine the polyion structure from the birefringence measurements in an electric field. Therefore, investigation of field-induced orientation of particles in dispersions is of primary importance for many areas of biophysics and biochemistry.

# REFERENCES


1. **Tolstoi, N. A. and Feofilov, P. P.,** On some electric phenomena in colloidal dispersions, *Dok. Akad. Nauk SSSR,* 66, 617, 1949, (in Russian).
2. **Kroit, G. P.,** *Colloid Science,* Vol. 1, Elsevier, Publishing Company, Amsterdam-Houston-New York-London, 1952.
3. **Van de Hulst H. C.,** *Light Scattering by Small Particles,* Wiley, New York; Chapman and Hall, London, 1957.
4. **Shifrin, K. S.,** *Introduction to the Ocean Optics,* Nauka, Leningrad, 1984, chap. 4 (in Russian).
5. **Kerker, M.,** *The Scattering of Light and Other Electromagnetic Radiations,* Academic Press, New York, 1969, chap. 1.
6. **Khlebtsov, N. G.,** *Light Scattering by Nonspherical Particles and Its Application to Turbidity and Electrooptic Studies,* Abstract of Ph. D. thesis, Saratov, 1979 (in Russian).
7. **Smirnov, A. V. and Shifrin, K. S.,** Light absorption by a dispersion, *Kolloid. Zh,* 49, 809, 1987 (in Russian).
8. **Shifrin, K. S. and Tuchkin, V. F.,** Reconstruction of size distribution of particles from the scattering diagram by statistical regularization, *Izv. Akad. Nauk SSSR, Fiz. Atmos. Okeana,* 8, 1268, 1972 (in Russian).
9. **Koch, A. L.,** Theory of angular dependence of light scattering by bacteria and similar size biological particles, *Zh. Teor. Biol.,* 18, 133, 1968.
10. **Ivanov, A. P.,** *The Optics of Scattering Media,* Nauka i Tekhnika, Minsk, 1969, chap. 2 (in Russian).
11. **Newton, P. G.,** *Scattering Theory of Waves and Particles,* McGraw-Hill, New York, 1966, chap. 1.
12. **Rozenberg, G. V.,** Electrodynamics of statistically inhomogeneous media, in *Theoretical and Applied Problems of Light Scattering,* Nauka i Tekhnika, Minsk, 1971, p. 159 (in Russian).
13. **Lind, A. C., Grenberg, J. M., Wang, R. T., and Libelio, L. F.,** *Electromagnetic Scattering by Nonspherical Systems,* Inst. Adv. Study, Princeton, New Jersey, 1966.
14. **Lind, A. C.,** Electromagnetic scattering by obliquely oriented cylinders, *J. Appl. Phys.,* 8, 3395, 1966.
15. **Mors, P. M. and Feshbakh, G.,** *Methods of Theoretical Physics,* McGraw-Hill Book Company Inc, New York-Toronto-London, 1953, chap. 4, p. 11.
16. **Sizikov, V. S. and Verlan', A. F.,** *Integral Equations: Methods, Algorithms, Programs.,* Naukova Dumka, Kiev, 1986 (in Russian).

17. **Nikol'skii, V. V. and Malushkov, G. D.,** The projection method in the problem of dielectric antenna excitation, *Izv. Vyssh. Uchebn. Zaved., Radiofiz.,* 16, 1045, 1973 (in Russian).
18. **Sirota, A. J. and Khlebtsov, N. G.,** Mechanisms of electrooptic effects in colloidal dispersions, in *Physics of Metal Strength and Electrodynamic Phenomena in a Substance,* Kuibyshev, 1976, p. 139 (in Russian).
19. **Stoylov, S. P.,** Colloid electrooptics. Electrically induced optical phenomena in dispersions, *Adv. Colloid Int. Sci.,* 3, 45, 1971.
20. **Stoylov, S. P., Shilov, V. N., Dukhin, S. S., Sokerov, S., and Petkanchin, I.,** *Electrooptics of Colloidal Dispersions,* Naukova Dumka, Kiev, 1977 (in Russian).
21. **Tolstoi, N. A., Spartakov, A. A., and Trusov, A. A.,** Electrooptic phenomena in a rotational electric field and the rigid dipole moment of colloidal particles, *Opt. Spektrosk.,* 19, 826, 1965 (in Russian).
22. **Tolstoi, N. A., Spartakov, A. A., and Trusov, A. A.,** Electrooptic properties of lyophobic colloidal solutions. 3. Techniques for studying the electrooptic effect in a rotational field. Foundations of theory, *Kolloid. Zh.,* 28, 735, 1966 (in Russian).
23. **Storonkin, B. A.,** Orientational relaxation in a strong dipole interaction induced by an external field, *Teor. Mat. Fiz.,* 41, 395, 1979 (in Russian).
24. **Storonkin, B. A. and Kuni, F. M.,** Rotational diffusion of dispersed particles in strong external fields, *Kolloid. Zh.,* 39, 1118, 1977 (in Russian).
25. **Storonkin, B. A., Melekhov, A. A., and Kuni, F. M.,** Electrical relaxation of dispersed particles in external fields, *Kolloid. Zh.,* 41, 289, 1979 (in Russian).
26. **Fedorova, I. S.,** Solution of inverse scattering problem in the Rayleigh-Debye approximation. 1. Determination of the diameter distribution of assembles of long cylinders, *J. Colloid Int. Sci.,* 59, 98, 1977.
27. **Khlebtsov, N. G., Schegolev, S. Yu., and Klenin, V. I.,** Account of particle specificity in determining their parameters from turbidity spectra. Characteristic functions of light scattering by soft flattened randomly oriented particles, *Opt. Spektrosk.,* 54, 563, 1978 (in Russian).
28. **Erugin, N. P.,** *A Reader in the General Course on Differential Equations,* Nauka i Tekhnika, Minsk, 1979, p. 408 (in Russian).
29. **Leontovich, M. A.,** *Introduction to Thermodynamics. Statistical Physics,* Nauka, Moscow, 1983, p. 378 (in Russian).
30. **Krylov, V. I. and Skoblya, N. S.,** *Methods of Approximated Fourier Transformation and Inversion of Laplace Transformation,* Nauka, Moscow, 1974, chap. 4 (in Russian).
31. **Philips, O. L.,** A technique for the numerical equations of the first kind, *J. Assoc. Comput. Mach.,* 9, 84, 1962.
32. **Tikhonov, A. N. and Arsenin, V. Ya.,** *Methods for Solving Non-Correct Problems,* Nauka, Moscow, 1979, chap. 4 (in Russian).

33. **Rafikov, S. P., Budtov, V. P., and Manakov, Yu. B.,** *Introduction to Physical Chemistry of Polymer Solutions,* Nauka, Moscow, 1978, chap. 2 (in Russian).
34. **Eskin, V. E.,** *Light Scattering by Polymer Solutions and Properties of Macromolecules,* Nauka, Leningrad, 1986, p. 76 (in Russian).
35. **Tarasevich, Yu. I. and Ovcharenko, F. D.,** *Adsorption on Clay Minerals,* Naukova Dumka, Kiev, 1975, chap. 3 (in Russian).
36. **Fedorov, V. A. and Aleksayan, V. I.,** Calculation of the length distribution function for cylinders from small-angle X-ray scattering, *Kristallografiya,* 15, 240, 1970 (in Russian).
37. **Fedorova, I. S. and Emelyanov, V. B.,** Solution of inverse problems of scattering in the Rayleigh–Debye approximation. 2. Determination of diameter distribution function of thin spherical shells, *J. Colloid Int. Sci.,* 59, 106, 1977.
38. **Frike, G. H., Rosental, D., and Welford, G.,** Fitting data with the *z*-distribution, *J. Phys. Chem.,* 74, 1139, 1970.
39. **Kato, M. and Eichinger, B. E.,** Small-angle X-ray scattering functions for aggregates of globular proteins, *Angew. Macromolec. Chem.,* 51, 61, 1976.
40. **Stevenson, A. F., Heller, W., and Wallach, M.,** Theoretical investigation on the light scattering of colloidal spheres. XI. Determination of size distribution curves from spectra of the scattering ratio or from depolarization spectra, *J. Chem. Phys.,* 34, 1789, 1961.
41. **Gledkill, R.,** Particle-size distribution determination by turbidimetry, *J. Phys. Chem.,* 66, 458, 1962.
42. **Wallach, M., Heller, W., and Stevenson, A.,** Theoretical investigation on the light scattering of colloidal spheres. XII. The determination of size distribution curves from turbidity spectra, *J. Chem. Phys.,* 34, 1796, 1961.
43. **Stoylov, S. P.,** Transient electric light scattering. 2. Determination of the distribution curves for solutions of polydisperse rods, *J. Colloid Int. Sci.,* 27, 542, 1968.
44. **Shifrin, K. S.,** *Light Scattering in a Turbid Medium,* Gostekhizdat, Moscow, Leningrad, 1951, chap. 4 (in Russian).
45. **Dukhin, S. S.,** *Electrical Conductivity and Electrokinetic Properties of Dispersions,* Naukova Dumka, Kiev, 1975 (in Russian).
46. **Dukhin, S. S. Estrela-L'opis, V. R., and Zholkovskii, E. K.,** *Electrosurface Phenomena and Electric Filtration,* Naukova Dumka, Kiev, 1985, chap. 2 (in Russian).
47. **Dukhin, S. S. and Shilov, V. N.,** *Dielectric Phenomena and a Double Layer in Dispersions and Polyelectrolytes,* Naukova Dumka, Kiev, 1972 (in Russian).
48. **Dukhin, S. S. and Deryagin, B. V.,** *Electrophoresis,* Nauka, Moscow, 1976, chap. 2 (in Russian).

49. **Fridrikhsberg, D. A.,** *Study of Charged Particle Movement in Capillary Systems and Their Electrokinetic Properties,* Abstract of post-Ph. D. thesis, Leningrad University, Leningrad, 1967 (in Russian).
50. **Martynov, G. A.,** Double layer in dilute electrolyte solutions. Discussion of a model and experiment, *Elektrokhimiya,* 15, 611, 1979 (in Russian).
51. **Eremova, Yu. Ya. and Shilov, V. N.,** Transverse polarizability of elongated spheroids in electrolytic. Solutions in a constant electric field, *Kolloid. Zh.,* 37, 635, 1975 (in Russian).
52. **Dukhin, S. S. and Mishuk, N. A.,** Strong concentrational polarization of a thin double layer of a spherical particle in an external electric field, *Kolloid. Zh.,* 50, 237, 1988 (in Russian).
53. **Dukhin, S. S., Mishuk, N. A., and Zholkovskii, E. K.,** Concentrational polarization of a thin double layer of a spherical particle at large Peclet numbers, *Kolloid. Zh.,* 49, 865, 1987 (in Russian).
54. **Malkin, E. S and Dukhin, S. S.,** Interaction of dispersed particles in an electric field and linear concentrational polarization of a double layer, *Kolloid. Zh.,* 44, 904, 1982 (in Russian).
55. **Simonova, T. S. and Dukhin, S. S.,** Nonlinear polarization of a thin diffuse double layer of spheres, *Kolloid. Zh.,* 38, 86, 1976 (in Russian).
56. **O'Konski, C. T.,** Electrical properties of macromolecules. Theory of ionic polarization in polyelectrolytes. *J. Phys. Chem.,* 64, 605, 1960.
57. **Rozen, Yu. Ya. and Shilov, V. N.,** Low-frequency dispersion of dielectric permittivity of rod-like particle suspensions, *Biofizika,* 19, 91, 1974 (in Russian).
58. **Malyarenko, V. V., Ovcharenko, D. F., and Dukhin, S. S.,** Study of electrical conductivity of dilute palygorskit suspensions and calculation of specific surface conductivity, *Kolloid. Zh.,* 36, 484, 1974 (in Russian).
59. **Sokerov, S., Petkanchin, I., and Stoylov, S.,** *Electrical Properties of Colloidal Solutions of Benzopurpurine in Collected Articles, Surface Forces in Thin Films and Dispersions,* Nauka, Moscow, 1972, p. 96 (in Russian).
60. **Ovcharenko, F. D., Malyarenko, V. V., and Shilov, V. N.,** Study of electrical conductivity of suspensions and electro-orientation of colloidal particles, *Kolloid. Zh.,* 39, 73, 1977 (in Russian).
61. **Kukovskii, E. G.,** *The Structure, Physical and Chemical Properties of Clayey Minerals,* Naukova Dumka, Kiev, 1966, p. 88, (in Russian).
62. **Stoylov, S. and Petkanchin, I.,** Effect of 1-1-valent electrolyte on electrical polarizability of colloidal particles in an aqueous solution, *Izv. Institute of Physical Chemistry, Bulgarian Acad. Sci.,* 5, 87, 1965 (in Bulgarian).
63. **Stoylov, S.,** Study of the electric polarizability of colloid particles by their light scattering, *in Proc. 4-th Int. Congress on Surface Activity,* 2, 171, 1964.

64. **Voitylov, V. V., Rudakova, E. V., and Trusov, A. A.,** Anisotropy of electric polarizability of diamond and anisaldazine in aqueous colloidal dispersions, *Kolloid. Zh.,* 40, 25, 1978 (in Russian).
65. **Eremova, Yu. Ya. and Shilov, V. N.,** Longitudinal polarizability of dielectrical spheroidal particles in constant electric field, *Kolloid. Zh.,* 37, 1090, 1977 (in Russian).
66. **Shilov, V. N.,** Thermodynamics of irreversible processes as applied to the theory of field-induced orientation effects in dispersions, *Dok. Akad. Nauk SSSR,* 200, 1161, 1971 (in Russian).
67. **Plummer, H. and Jennings, B. R.,** Light scattering by rod-like macromolecules in alternating electric fields, *J. Chem. Phys.,* 50, 1033, 1969.
68. **Grilikhes, M. S. and Filanovskii, V. K.,** *Contact Conductometry,* Nauka, Leningrad, 1980, chap. 4 (in Russian).
69. **Lopatin, B. A.,** *Conductometry,* Novosibirsk, 1964, 21, (in Russian).
70. **Filanovskii, B. K. and Grilikhes, M. S.,** Polarization phenomena in contact conductometry. 4. Types of equivalent circuits and cell constants, *Zh. Fiz. Khim.,* 52, 1076, 1978 (in Russian).
71. **Zharkikh, N. I. and Borkovskaya, Yu. B.,** Theory of nonequilibrium electro-surface phenomena in concentrated dispersions. 2. Calculation of kinetic coefficients in dispersions of thin layer cylinders, *Kolloid. Zh.,* 43, 652, 1981 (in Russian).
72. **Shilov, V. N., Zharkikh, N. I., and Borkovskaya, Yu. B.,** Theory of nonequilibrium electro-surface phenomena in concentrated dispersions. I. Thermodynamics of irreversible processes as applied to a cell model of concentrated dispersion, *Kolloid. Zh.,* 43, 540, 1981 (in Russian).
73. **Borkovskaya, Yu. B., Zharkikh, N. I., and Dudina, J. M.,** Electrical conductivity of concentrated dispersions in thin double layer approximation, *Kolloid. Zh.,* 44, 645, 1982 (in Russian).
74. **Voitylov, V. V., Kakorin, S. A., and Trusov, A. A.,** Study of electrical conductivity anisotropy of colloids and suspensions induced by an external electric field in colloidal dispersions. *Kolloid. Zh.,* 48, 139, 1986 (in Russian).
75. **Schwarz, G.,** Zur Theorie der Leitfähigkeits Anisotropie von Polyelectrolyten in Lösung, *Z. Phys.,* 145. 563, 1956.
76. **Baran, A. A. and Sodoshentseva, I. I.,** Temperature dependence of stability of hydrophobic salts containing water solvable polymers, *Kolloid. Zh.,* 36, 6, 1974 (in Russian).
77. **Deryagin, B. V.,** *Theory of Stability of Dispersions and Thin Films,* Nauka, Moscow, 1986 (in Russian).
78. **Ovcharenko, F. D., Kukovskii, E. G., and Nichiporenko, S. P.,** *The Colloidal Chemistry of Palygorskit,* Naukova Dumka, Kiev, 1968, chap. 1 (in Russian).
79. **Pilipenko, A. T., Falyndyts, I. F., and Parkhomenko, E. P.,** The state of 3-aluminium in aqueous solutions, *Khimiya i Tekhnologiya Vody,* 4, 136, 1982 (in Russian).

80. **Matijevich, E.,** Detection of metal ions hydrolysis by coagulation 3-aluminum, *J. Colloid Chem.,* 65, 826, 1961.
81. **Matijevich, E. and Strejker, L.,** Coagulation and reversal of charge of lyophobic colloids by hydrolyzed metal ions of 3-aluminum sulfate, *J. Colloid Sci.,* 22, 68, 1966.
82. **Dudkina, L. M. and Baran, A. A.,** The effect of adsorbed polymers on the surface conductivity of colloidal particles, *Teor. Eksp. Khim.,* 15, 162, 1979 (in Russian).
83. **Dukhin, S. S. and Dudkina, L. M.,** The effect of adsorption of non-coherent molecules on electrical conductivity of suspensions, *Kolloid. Zh.,* 40, 232, 1978 (in Russian).
84. **Radeva, Ts., Baran, A. A., and Dukhin, S. S.,** Conductometric and electrooptic studies of rod-like palygorskit double layer, *Kolloid. Zh.,* 42, 685, 1980 (in Russian).
85. **Chardymskaya, E. Yu., Sidorova, M. P., and Fridrikhsberg, D. A.,** Electrical conductivity of quartz suspensions in 1:1 charged electrolytic solutions, *Kolloid. Zh.,* 48, 589, 1986 (in Russian).
86. **Wirsema, P. H., Leeb, A. L., and Overbeck, J. T.,** Calculation of the electrophoretic mobility of a spherical colloidal particle, *J. Colloid Int. Sci.,* 22, 78, 1966.
87. **Hunter, R. J.,** *Zeta-Potential in Colloid Science. Principles and Applications,* Australia Academic Press, Sidney, 1967, chap. 1.
88. **Deryagin, B. V., Churaev, N. V., and Muller, V. M.,** *Surface Forces,* Nauka, Moscow, 1985, chap. 7 (in Russian).
89. **Semina, L. A.,** *Comparison of Electrokinetic Properties of Ionic Crystals and Oxides in Aqueous Electrolytic Solutions of Barium Sulfate and Quartz,* Abstract of Ph. D. thesis, Leningrad University, Leningrad, 1976 (in Russian).
90. **Sidorova, M. P., Dmitrieva, I. B., and Golub, T. P.,** Investigations of electro-surface properties of quartz in 1:1 electrolytic solutions, *Kolloid. Zh.,* 41, 488, 1979 (in Russian).
91. **Fazylova, M.,** *Study of Zeta-potential and Surface Conductivity of Quartz Fibers in Electrolytic Solutions,* Abstract of Ph. D. thesis, Leningrad University, Leningrad, 1976 (in Russian).
92. **Niskael, H. L. and Williams, D. J. A.,** Electrochemical properties of quartz, *J. Electr. Chem.,* 179, 131, 1984.

# INDEX